CUT
- & -
CRUSHED

by Frank H. Edwin

Edited and contributions by Raichelle de Caan

The Old School Agricultural Engineer's Guide to
Surviving the Tropics -
and still making it home for Christmas

HD
HaraldsDottir

Published in 2025 by HaraldsDottir

HD
HaraldsDottir

ISBN: 978-1-913898-95-3
Also available as an ebook

Cover and Book interior Design by Russell Holden
www.pixeltweakspublications.com
Pixel Tweaks Publications
SELF PUBLISHING MADE SIMPLE

The events and conversations in this work are accurate to the best of the author's, editor's and contributor's knowledge and recollections, sources and research. However, bar a few senior figures, commercial enterprises and geographical locations, whose acts and existence are all matters of reasonably-accessible public record, all names have been changed or removed to protect the innocent; or guilty, as the case may be.

For Dorothy

With love and thanks

CONTENTS

BEFORE THE BEGINNING

I have been a stranger in a strange land.
Exodus 2:22

Who could possibly have imagined, back at the start of the twentieth century..?

Oh. No. Someone else already used that, didn't they?

All right, then. How could I, as a young boy, born in the industrial heartlands of England, just after the end of the First World War..?

Well, you get the point.

The fact is, life, in all of its facets, has changed almost beyond recognition over the last hundred years. My working life - a very good part of it, at least - belonged to an age well before the pocket calculator, never mind the computer, the internet, the smartphone, and whatever other systems and gadgets might be in common use by the time that you're reading this. Just a basic, reliable electricity supply was often far from guaranteed in those far-flung, oft-tropical parts of the world where I plied my trade.

Let's get it clear from the start, though: whilst most of the countries which I visited did indeed fall well within the general geographical categorisation of 'the Tropics', those now well-associated words like luxury, paradise, palm-fringed, beachfront, and their ilk somehow - for me, at least - rarely seem to have been involved. Instead, most of the places to which I went were, at the time, much more usually referred to as part of the 'third world', or, at best, 'developing nations', mostly unheard of, frequently isolated, and usually very sparsely populated.

Nowadays, of course, many can be encountered with ease,

advertised as exotic tourist hotspots or 'eco-friendly' adventure destinations, each filled with national parks, heritage centres, and all kinds of other attractions. Some have simply expanded into, or been taken over by, large, thriving towns and commercial cities, their multi-lane highways serving sprawling hotels, shopping malls, and industrial parks.

Or maybe, sometimes, sadly lying in the ruined tatters of conflict and war.

Transportation, in all of its forms, was also still very much 'developing', and definitely different. Efficient, on-demand, and certainly instant communications were, to a good extent, simply non-existent. There were no satellite or mobile phones, of course: a single landline telephone and, later, a fax - both pretty much things of the past, now - was, in many cases, the most that one could dream of, and often no more than a dream. And as for 24/7 voicemails, e-mails, video calls, photo and document sharing, texts, messaging, GPS location devices, or easily-portable radios..? I can only envy the ease with which my modern compatriots can - usually, I'd better hastily qualify - go about their business, wherever and whenever they happen to be.

For me, except in the seriously-direst of emergencies, when employer, consular or embassy assistance might have been summoned - itself necessitating some form of oft-long and difficult drive to the nearest radio or fixed telephone line - my only reasonably-speedy form of contact with the outside world was by means of the costly 'telex', where any message was paid for by the precious word, and thus rarely consisted of anything more complex than "contact home" - or office - "soonest". Fortunately, most larger commercial premises, post offices, or telephone exchanges in the more major areas of civilisation would have access to at least one of these globally-networked teleprinter-type devices, and would thus be able to dispatch, and in turn helpfully receive, even deliver, such messages as might be required. When not in such a location, of course - and I frequently wasn't - a good drive would again be required, combined with the swift making of amicable arrangements with the nearest person possessing a radio, or perhaps an aircraft, and thus able to pass or carry urgent messages up and down through the appropriate chain.

In all other cases, the exchange of any other information, not only with my UK and worldwide colleagues and business contacts, but my wife and family back in England, was by means either of a small postcard, clearly marked and paid for as 'via airmail', or the more common prepaid, standard-sized, lightweight, fold and lick, blue paper 'airmail letter', within which, in your practised tiniest handwriting, you learnt to convey - and eventually received back, a couple of weeks or so later if you were lucky - all of your news. This, of course, included not just your worries, concerns, questions, answers, advice - technical as well as familial - and general personal endearments, but the plain, simple facts of safe arrival, current and future location, and planned return dates and arrangements. Assuming, of course, that someone ventured out with any frequency to the nearest postal delivery and collection point.

Drawings, documents, reports and the like, all printed or typed out on reams of heavy paper, simply had to await some convenient person to transport them in person when they next happened to fly out or back.

Educating said spouse, offspring, and any other interested friends and colleagues as to where I'd been and what I'd been getting up to - both by now firmly past tense - also had to await not only my return home, but the eventual slow development of whatever 24- or 36-exposure rolls of 35mm slide (and later print) film I'd been able to shoot whilst there: assuming, of course, that it hadn't by then got 'fogged', and thus rendered useless forever, by the multitude of airport X-ray scanners through which it had been required to pass on my various outbound, in-country, and return journeys, or just ruined by the general tropical heat and damp.

The politics, customs, and accepted mores of my times were very different in so many ways, too; although, upon reconsideration as regards a few certain ethics and morals, perhaps not always so much.

Well, politics maybe not always so much either, come to think of it.

But both working practices and people were definitely not what you'd be used to today; and neither were accepted gender roles and their accompanying stereotypes.

Varying countries' entire cultures, governments, and systems will have changed vastly by now. Several are now known by entirely different names. A few might not even exist.

All in all, therefore, I suspect that some of you - perhaps most of you - will have to employ more than a little imagination, not to say suspend a few modern preconceptions, in the face of what was, in those days, my reality.

But that's part of what stories of the past are all about, isn't it?

Much of my overseas working life was, to me, repetitive and routine. Some, however, was most definitely challenging, physically and otherwise. There were times, at least with hindsight, when it was certainly amusing, if not utterly downright bizarre; and times when it most definitely wasn't funny at all. Some incidents, with yet more of that wonderful retrospection, were frankly nothing but downright dangerous; or, at least, stemmed from stupidity in the extreme. But all of it left memories: of the sweet times and the sour, of good times and good friends, and of nothing but the irritations, annoyances, and frequent frustrations of all too oft-encountered problems.

Fortunately, even the worst experience can ultimately - if not at the time - now be recalled and looked back upon with a reasonable degree of enjoyment.

Mostly.

It is not my intention to describe my travels in detail, nor the scenery, history, or peoples of the countries to which I went: you can find that information, and all kinds of accompanying photos, in plenty of places by now, I'm sure. Neither do I care to bore you with the intricacies and technicalities of my work, never mind the many and varied in-office and on-site interactions and decisions which my day to day life entailed: though I hope that you'll pick up a flavour from time to time.

I'm certainly not attempting to make any particular points, either, political or otherwise; and, as far as I'm concerned, any debate concerning the meaning, interpretation or impact of any matters, current or historical, as they relate to the particular industry in which I was employed is, as they now say, way above my pay grade. I don't, and never did, claim to be an expert on the politics, national or local, of any of the places which I visited, either. In fact, I always rather wished to know as little of them as possible, and we all, company and personnel, always tried very hard to keep ourselves completely out of them, except to the practical and pragmatic extent that they directly involved or impacted upon our duties, our employer's or contractors' commercial and operational interests, or the safety of ourselves, our staff and workers. Hopefully, or so the thought process went - and, to a very good degree, usually seemed to work well enough in practice - any related disputes and clashes then didn't get in our way by return.

This is therefore nothing but a collection of events as I remember them, occasions as I personally experienced them, anecdotes about people who, and events which, left such impressions on me that I still carry them with me decades after I retired. It is merely, from a very personal point of view, a flavour of what my work, life, and our wider world was like back then, at - and only at - the time when I was there.

I intend no specific offence to anyone, nor their homelands, beliefs or cultures, and I sincerely hope that none is taken.

Finally, I should make it clear that, simply in order to correlate things into some sort of pattern, I have grouped them by the countries and regions in which they occurred, and not in accordance with any form of time sequence at all. What is described in one chapter may therefore have taken place over a long period of time, and will, like as not, have overlapped or run parallel with many other related events taking place in a completely different part of the world.

Such, also, was how it happened.

1

ACORNS AND AFTERWARDS

It's been an interesting job. I've flown all over the globe, spent a fair amount of time hanging around a great number of international airports, and run full pelt through quite a few of them, too. I've missed planes, caught them by the skin of my teeth, even personally flagged them down as they taxied out towards the runway; and I've certainly taken off and landed on a good number of exceptionally-questionably shaped and contoured ones of those as well. I've sat for days in aircraft spewing flames and oil, literally scraped over treetops and mountain peaks, survived storms of both torrential tropical waters and thick desert sands, examined the insides of an active volcano at very close quarters, and nearly become lost forever in the middle of the Amazonian jungle. I've arrived at gunpoint in the middle of an armed gang dispute, been tracked and nearly shot down by surface-to-air missiles, acted as an impromptu government spotter tasked with monitoring a revolutionary uprising, and formed the apparent aerial vanguard of a long-feared invasion force.

Once safely back on the ground, I've explored and plotted the most remote of uncharted lands: places which few - at the time, if not necessarily now - had even heard of, and even fewer - the usual local terrorist insurgents excepted - would ever have wanted to go. I've discovered forbidden voodoo temples, hacked and heaved my way up virgin mountain slopes, and then carried collapsed colleagues back down them again. I've ended up lost and marooned just about everywhere from rainforests to dry swamps, never mind the endless varieties, and even more endless swathes, of thick, impenetrable bush, grasslands, and scrub. I've demonstrated my prowess at all kinds of extreme and off-road driving, skidding along surely every possible type and colour of slippery surface, plunging bonnet-first through raging torrents and across white-water weirs, traversing old, long-deserted rail beds for hour upon desperate hour in search of any sign of civilisation at all. I've sampled the delights of just about every kind of swinging, swaying, and decidedly-questionably-supported, if not in reality wholly-structurally-unsound, so-called bridge - not to mention their mandatorily-accompanying deep ravines, filled with rushing, rain-swollen rivers - than I ever fully care to consider again. I've fished for my supper over the edge of waterfalls the size of Victoria Falls, hand-build makeshift vehicular crossings in the dead of a pitch-black night, and generally faced more tests of cunning, problem-solving, ingenuity, initiative, and self-sufficiency than your modern-day reality TV show could ever dream of.

I've been smuggled over closed international borders, rescued compatriots from the clutches of kidnappers, stood up to threats of extortion, had government officials hold up entire transatlantic jets pending my late arrival, and couriered substantial wads of extremely-suspect cash between a few places which I probably really shouldn't ever have mentioned further. My employers and associates were publicly denounced as the pushers, promoters, and purveyors of one of modern life's greatest evils and hazards to all human health. My fingerprints had to be lodged at Interpol. Once, having narrowly failed to electrocute me, someone even tried to blow me up.

I also recall spending not an inconsiderable amount of time surrounded by - and, of course, fending off the advances of - countless young, glamorous, and decidedly-exotic women.

One way or another, though, I've definitely not always been the maker of the best first impressions. I've been detained as an illegal immigrant on more than one occasion - perhaps unsurprisingly, you might by now be thinking - and, once they did eventually let me in, quite a few countries then denied me permission to leave again. One even decided to hold me for ransom. But at the end of the day, thankfully, I did always somehow manage to make it safely back home, with nothing more exciting to show for it than a scarred lip and a few shop-bought souvenirs.

I was also, occasionally, regarded as somewhat of an expert in my field.

For those of you now wondering exactly what field that was, if they've inadvertently picked up the memoirs of some kind of jet-setting international spy, a covert intelligence or special forces operative - not, of course, that I'd either confirm, deny or generally tell you if I was, would I? - an intrepid journalist, politician, diplomat or the like... Then the answer's no.

I did admittedly buy and read an awful lot of novels about them all; and I've certainly had to plot and negotiate my way out of more than a few sticky situations. But, my fair share of scrapes and lucky escapes aside, mine was a much more mundane and practical role.

Although sticky, come to think of it, might still be a very apt word: for that was indeed my employer's main business, I being its mere overseas representative, a consultant selling his services in the very literal field of the cultivation, harvesting, transporting, and processing of sugar cane.

So there. That's it. Nothing exciting at all. All quite boring, really.

And yet... Well, yes: I suppose that it did have its moments.

It's more than fair to say that at the appointed time for my leaving school - January 1936, my sixteenth birthday - it wasn't exactly the life which I was expecting. On the other hand, a few unforeseen 'modern' developments and their attendant requirements and consequences aside, it was also exactly the life for which I'd hoped.

My formative years, between the First and Second World Wars, began in the residential streets of inner-city Derby, the industrial railway hub of the English Midlands; they continued in its newly-expanding, if then still-rural, outskirts. This wasn't the ancestral home for one half of the family, though, my paternal grandfather having previously undertaken an unusual, perhaps trend-setting, career-changing relocation from Bedfordshire, where generations of forebears, and now his siblings, had all worked as servants and labourers on the Duke of Bedford's estate. Nevertheless, it was a city, I'd assumed, where I would be spending the rest of my days.

My family had certainly moved itself a little up the social scale by the time of my arrival. We definitely weren't well-off, but we weren't poor, either; for the most part, anyway. Our close relations encompassed a variety of builders, several lace makers, various skilled railway workers, one train driver, a very competent businesswoman-cum-trailblazing-DIY-er, and a good selection of mechanical engineers. All of these were considered respectable positions in the general working people's scheme of the day, and we therefore enjoyed a reasonably comfortable standard of living: restricted food supplies, mounting unemployment, and all kinds of post-war financial, economic, and assorted other pressures notwithstanding.

Admittedly, our domestic living accommodation was, at times, a little more than crowded, my first five years being shared not only with my parents, but with my mother's parents, her younger brother, and a young girl who was in reality some form of more distant cousin, but had been adopted by my grandparents in one of those embarrassing family situations which really wasn't talked about then; or, indeed, for a very long time after. And that was all before my baby sister came along: eight of us in total, all crammed into a cold, two-bedroomed terraced house, sharing one living-room (there were, in fact, two, but the other was only ever opened up 'for best'), one outside 'privy', and one tin bath in front of the fire. This wasn't the fault of financial circumstance, however: just the terrible housing shortage which existed after the First World War.

Like our peers, we all lived very much by the 'make do and mend' ethos of the times, but still enjoyed a good mix of music, reading, and lively discussion around a very wide range of general interests; and, of course, a great many social, educational, and other activities centred around our respective local churches, in turn an odd mix of staunch Anglicans, strong Primitive Methodists, and, more latterly, very tolerant Baptists.

We also, for the time, seemed to have possessed an unusual thirst for travel, enjoying every opportunity for local walks and other outdoor journeys of discovery, including frequent visits to friends and relatives in whatever nearby vicinity one could reach and return from in a day. Every possible combination of tram, omnibus, local train, and foot - there were no personally-owned cars for us, back then - would be employed to aid our expeditions. The bicycle, however, was actually the one mode of transport which we didn't use, the only one in our family belonging firmly to Dad, and being the highly-valued, and thus highly-protected, means by which he commuted back and forth to his daily work. We also always managed to take at least one annual week's, sometimes fortnight's, holiday, sampling a good selection of whatever far-distant seaside resorts the national rail services could carry us to. But wanderlust must have been in our genes, for even back then, I later discovered, we already had relatives who'd ventured to live not only in another district of Derby - in itself a far-from common occurrence of the day - but had actually left the country itself, emigrating to the distant realms of Canada, the USA, and New Zealand. We never, of course, got to visit them, nor they us: but both my mother and father also had uncles and cousins living and working in far-flung English towns and cities throughout Yorkshire, Derbyshire, Staffordshire, and Bedfordshire, occasioning still quite long-enough-distance family visits, train and bus routes permitting.

Aided by many past family attendees, and more than a little parental determination, a place was eventually secured for me at a reasonably-nearby, but very highly-regarded primary school. (Some things, it seems, are not quite as modern a phenomenon as one might sometimes imagine.) From there, purely under my own efforts, I managed to pass both the general 'eleven-plus' and a respected grammar school's own entrance exams, gaining a County scholarship in the process. But aspirations towards any further, certainly university-based, education were not for the likes of me. The best, and most obvious route forward which I could hope for would be to gain an apprenticeship, attend a local college, and eventually attain a sufficiently-respectable position 'in industry'.

Grandfather, having broken away from the traditional chain of servitude, had initially joined his local railway company as a porter. He must have impressed someone, as he quickly managed to work his way up through various positions, eventually transferring, via Somerset, to Derby, where he retired as head signalman. Three of his four sons had duly followed him into the new family business, all becoming skilled artisans at the Midland Railway Company's locomotive works: one an electrician, one a plate bender and riveter, one a tool maker. My father, however, come his fourteenth birthday (the then official school-leaving age), remained disappointingly small and weak. Forced to seek alternative, more physically-appropriate employment, he'd found a position at one George Fletcher & Sons Ltd, manufacturers of sugar-industry machinery, and also the workplace of my future mother's brother: such facts, of course, along with subsequent social interactions, contributing in no insignificant part to the eventual cause of my existence.

Some sixteen years after that happy event, and now approaching my own designated transition into the world of work, I had already firmly decided that I wanted to follow in the footsteps of my father and uncle - at that time both working in the 'drawing office' as, respectively, an engineering estimator and the company's specialist mill designer - and train to become a 'draughtsman'. Long before taking my School Certificate examinations, I had painstakingly written out, by hand, to all of the various businesses in the Derby area which had their own in-house engineering offices, enquiring as to any likely forthcoming vacancies. But to my dismay, although fast approaching the end of my final school term, and just two weeks before Christmas, no replies, never mind offers, had yet been received.

Then a local electrical contractor, Wallace Cockburn, approached the school. His last apprentice, he explained, had been one of their scholars, and had proved very satisfactory indeed. Was there, therefore, another suitable boy who might like to join him?

Straightaway, my form master sent me down to be interviewed - more of a command than an invitation, it felt - and I was immediately offered the position, starting the following Monday.

Bang went my last Christmas holidays.

My stint as an electrician only lasted four months. This wasn't my employer's fault; nor mine. I was, in fact, thoroughly enjoying my new life, and had already learned a great deal - including a whole new vocabulary, mostly comprised of sexually-connected swear words - helping out with installations in an oil refinery, the county's mental hospital, a large sawmill, and several private dwellings. (The oil refinery, incidentally, burned down some years later, apparently due to an electrical fault.)

And then, unexpectedly, a letter arrived, asking me to attend an interview with the aforementioned 'Fletchers', who now had an opening for a drawing office apprentice. I naturally jumped at the chance; not to mention the princely sum of twelve shillings and sixpence a week. The training there was renowned throughout Derby, since their range of machinery and equipment was so very varied in type and design. It was also the usual order of things for young men, once having served their initial time and learned their basic trade, to seek further experience in other companies; and anyone successfully apprenticed at Fletchers could surely anticipate being offered future employment almost anywhere that he chose. (For the record, by the way, this was not an opportunity at all instigated by either my uncle or father - in fact, the latter was initially very much against the idea, and it took some serious persuading to allow me to accept the post - but by one of Dad's superiors, the most senior designer in the drawing office, who must have thought that, between my enthusiastic interest and my two family members' well-regarded abilities, I might just be the natural choice. His judgement can't have been that far off, either: many years later, once he'd retired, I ended up taking his job.)

So Wallace Cockburn and I parted ways, still on very good terms, and without any resentment: he'd always known where my real ambitions lay. I owe a lot to Wally and his men, though, for, even in that short time, they'd already taught me so much which, in both my personal and professional roles, would later stand me in very good stead.

And I don't just mean the bad language.

Having commenced my new in-house training, I also enrolled at Derby College of Technology, happy to discover that, having already attained 'matriculation standard' in both mathematics and physics, I was exempted from the entire first year of its course in engineering. This, therefore, in addition to my full-time day work, left me with only four more years of evening attendances, three per week, plus six hours of weekly homework, and many, many end of term examinations to look forward to.

I started, as all new boys did, in the factory's print room, making up the (literally-coloured) blueprints and copies of sketches and specifications; and, of course, also acting as their general dogsbody, answering the phone and running all internal errands required. It was usual for an apprentice to spend about a year here before moving out into the design office proper, where about forty 'drawing boards' were set up, and upon which, with the aid of various fine pens and pencils, solid-shaped and adjustable 'drawing instruments', slide rules, and huge tomes of tables and pre-calculated specifications, the company's products were designed. I flatter myself, however, in thinking that I was so keen in completing my assigned tasks, as well as somehow finding time to do some additional engineering sketches of my own, that I was promoted into the main office after only six months; co-incident, as it happened, with the early arrival of another newly-apprenticed lad.

The work at Fletchers was often quite seasonal, since orders for sugar factory parts and spares mostly arrived in line with the cane-crushing months of the countries where it grew. We were therefore extremely busy at some times of the year, but, during others, sometimes had very little to do. This, however, was fine by me and my fellow trainees, not only allowing us to catch up with our studies - the management were kind enough to allow their apprentice engineers, usually five or six of us at any time, one per year, to do their 'tech' homework in the office, once our normal duties had been dispensed with - but giving us time to go down into the main works and watch what we'd previously only seen on drawings being made into actual machines.

There was, of course, also time for the devil to find work for a group of idle young engineering minds. From initial paperclip-flicking trials and investigations into the glide path of paper aeroplanes, we proceeded into studying the more advanced realms of, for instance, the effects of dry ice upon partially-filled desk inkwells (messy is the obvious answer), and the psychological effects upon older co-workers of pouring water, clips, ink, and glue down through hollowed-out knots in the wooden flooring above them (interesting, we'll just leave that one as).

In June 1940, aged twenty, clutching one newly-awarded Higher National Certificate, my apprenticeship as a design engineer was officially complete. Somewhere along the line, squeezed in-between all of my work and studies, not to mention her own early and late shift work, I'd managed to continue 'walking out', and then more seriously and successfully to 'court', a lovely girl. Eleanor and I had first met a few years beforehand. Now, we become engaged. The wedding should have waited a few more years, at least until I'd settled myself into my career and she'd attained the age of legal majority, twenty one. But as it was - and nothing to do with the usual reasons, but everything to do with external events and forcibly-relocated parents - we were actually married, on rather short notice, within twelve months.

By now, of course, we were all well into the early throes of the Second World War. I'd already twice been 'called-up' to join the armed services; and twice failed the medical examination, due to my extreme short-sightedness. On the second occasion, however, the Ministry of Labour also designated me as being of a specialist 'reserved occupation'. Accordingly, perhaps also considering other experience which I'd already gained of working both on marine engines for naval corvettes and some new fangled types of rocket projectiles - it wasn't always dull sugar-industry machinery - it wasn't long before I was told to report to another Derby-based company, Coles the crane-makers, to carry out essential, military-orientated work.

I soon found myself part of a busy team of eight young designers, with an equally young chief engineer, charged with adapting and developing the firm's existing mobile range to meet various military needs. We turned out on average one unit per day, and I'm pretty sure that by the time the war ended there wasn't a naval depot, army maintenance unit or, particularly, airfield without one or more of our cranes.

Our team's first unexpected - and totally untrained and unprepared for - discovery was that, within a very short timeframe, we'd have to rewrite all of the firm's standard operating and maintenance manuals. They simply weren't easy enough to read and understand by those non-technically-trained members of the forces who'd nevertheless have to use and repair whatever we designed and built for them. We also had to prepare descriptions and illustrations of all the various spare parts which might be required, again in such a way, and using appropriately language and layouts, as to make their identification and fitting utterly and unambiguously obvious. As technical engineers, we weren't anywhere near capable of preparing the necessary quality of finished drawings: but we soon became competent at making good-enough rough pictorial, often three-dimensional sketches to enable an external art studio to produce the required final results. Mike Miller, it turned out, was also a keen photographer, so he was further able to assist in illustrating our new layman's instructions, his black and white prints being sent away for air-brush 'touching up' before being transformed into printing blocks. We then wrote up the text and parts lists between us, and arranged for the whole lot to be printed up into the short booklets which we'd devised.

Wartime certainly presented us with many proper engineering challenges, too. It was amazing how many variations we had to make to the long-proven, tried and tested crane designs in order to overcome the many and various ways which assorted service personnel managed to bend, break, and generally abuse them: ramming bridges, jamming them under arches, picking up loads either too great or beyond the outreach of all design limits, and generally attempting anything and everything hitherto unforeseeable to test our ingenuity - and patience - to its limits.

Then there were the increasing numbers of odd, often initially quite impossible, government orders which we soon started to receive, requesting us to design, fabricate, adapt, and modify - and waterproof, to fully-submersible standards - all kinds of rail- and truck-mounted lifting equipment, along with their engines and associated transports. More and more members of the three main British and US armed forces started arriving at our factory, and our strange products' development and testing phases were soon accompanied by a veritable posse of uniformed, if otherwise unspecified and utterly unidentified, 'experts'. As soon as they were completed, each individual item was then promptly whisked away, in person, by members of whichever service they'd been allocated to. Yet somehow, it never really occurred to us question the purpose of this undoubtedly experimental work; nor to think of it as having any great importance. It was just, we assumed, someone's strange idea, some typical eccentric, idiosyncratic scientist's hare-brained scheme, some desperate invention at last being allowed through into the prototype phase. And who were we to argue?

It was only after the 1943 Italy campaign, and the later Normandy D-Day landings, when we finally realised the significance of what we'd been doing, as the cinemas' newsreels revealed the numerous equipment-laden vehicles being taken ashore through the shallow sea waters from their landing ships. It also made us aware, in retrospect, of just how many similar preparations must have been needed, and undertaken, in advance of those operations, and the secrecy and breadth of planning required for them to take place.

We were usually called upon to test and demonstrate whatever we'd proposed, too. This led to many highly amusing, and a few downright dangerous incidents: but it also meant that I had to learn to drive. So I obtained a wartime 'essential use permit', and gained my basic training from a few helpful colleagues, a thirteen ton Thornycroft chassis, and a few turns around the factory yards in the works' sole 8-horsepower car. (Come the war's end, my temporary, restricted licence somehow found itself being converted under some time-saving procedure or other into a full, all-vehicle, road one: so I never did have to pass any kind of driving test at all. Which, no, doesn't explain a single thing at all; honest.)

My assigned wartime duties most definitely didn't end there. Prior to, but clearly in anticipation of, the outbreak of WW2, my father had trained to become an ARP (air raid precautions) warden, the front-line font of all local knowledge, instruction, and advice for the civilian population, and initial provider of on-site assessment, assistance, and overall co-ordination should the worst happen; and, of course, as all 'Dad's Army' fans will know, the much-maligned volunteer who had to patrol the streets at night telling everybody to close their blackout curtains and put out their lights. I, in the meantime, had been recruited into the Fletchers firewatch team, ready to leap into action - armed with one stirrup pump, one bucket of water, one bag of sand, and not much else apart from hope and good intentions - should we have to deal with the highly-feared incendiary attacks. As soon as I transferred to Coles, however, combining knowledge gained through my father's post with that of my own, I was put on charge of all of their civil defence matters, including warden, fire, and first aid services not only for our factory, but for the large food-bottling factory next door, complete with its raging glass-making furnaces and all.

It wasn't long before I also become a warden, taking over Dad's post when he was suddenly relocated to London to work at the Distillers Company, assisting the war effort by heading up a new engineering division, and a team designing equipment to produce explosives from imported molasses; and then, when that supply was restricted by the Japanese entering the fray, those new wonder chemicals, penicillin and sulphonamides. (And there he remained after the war, eventually retiring as chief engineer for the whole of their group's non-alcohol holdings.) Soon thereafter, the authorities invited me to train up as a warden instructor for the renamed Civil Defence Service. We were taught to handle all possible scenarios, from gas attacks, extinguishing fires, incident and traffic control, rescuing people from burning buildings and crew from crashed aircraft, rendering emergency first aid, and dealing with any childbirths which might also just happen to come along at the same time, our ultimate objective then being to pass on all of this accrued information in weekly training sessions at our area's wardens' posts.

The Coles factory, like so many others, also had its own Local Defence Volunteer platoon (or Home Guard, as it was soon known), ably headed-up by the chief engineer, and, yes, containing just about every character, and carrying out just about every exploit, seen during the TV series. Several fringe members, including myself and many of my in-house and external CD teams, quickly became unofficially attached in some capacity or other. We also soon developed a good working relationship with the local regular army's platoon instructor, for whom we manufactured many an odd, but obviously-helpful thing in our spare time. In return, no questions necessarily asked, he sent us back all manner of interesting items which he'd somehow happened to acquire - a few glass 'sticky bombs', a whole box of assorted grenades, and a 9mm German Luger automatic pistol complete with ammunition - all of which, when not actively being examined or practised with, fitted quite nicely into an easily-accessible design office cupboard, alongside our general rifles. The first of the new, and newly-dropped, German anti-personnel 'butterfly bombs' which our friendly sergeant cautiously carried in one morning - for some unknown reason, which he'd rather like to understand as soon as possible, as yet unexploded - I just examined and defused on the top of my desk.

We achieved a surprising amount of live-fire experience, too. In fact, I reckon that I very quickly learned to use just about every firearm in the then infantry book. Our factory was next door to an old brickwork's yard, once used as summer tennis courts and a winter ice rink, but now serving as a firing range for the nearby 'regulars' depot. They, however, only used it once a week, so for the other six days it was ours to do with as we liked. Our lunchtimes were often passed happily blasting away not only on the standard .303 rifle, but with our several Sten machine guns, the Luger, a Northover Projector - a simple canon designed to fire 3-inch diameter incendiary bottles filled with petrol and phosphorous - and a Boyes anti-tank rifle, the latter discharging a solid steel bullet of sufficient calibre that, even when fired prone, it could lift one bodily off the ground and several inches backwards. (In the current colloquial style of writing, you may here insert a very widely-grinning emoji, if you so choose.) We were also, as part of the Home Guard, thoroughly encouraged to make up and test out our own weapons: which we all duly, and to varying degrees of practical success, did. Quite surprisingly, considering the curiosity, creativity, and capacity of your average eager young design engineer, my first aid skills were never once required.

My initial well-intentioned efforts at remedying the factory CD team's lack of non-classroom training probably also fell under the same general heading of over-enthusiastic. With the approval and aid of both factories' managements, we prepared for the not-so-proverbial bomb to drop. In order to make the exercise as realistic as possible, I'd managed to obtain a superbly-loud thunderflash, along with quite a few smoke canisters, the latter to be lit by willing helpers as soon as the alarm sirens went off, and the staff tried to make their way to the shelters. I deliberately placed all of the smoke bombs only within the Coles factory, so as not to affect the valuable food-producing processes next door. But whether I'd placed more than I should, or whether this particular batch were much more vigorous than I'd imagined, I don't really know; a certain oversight in terms of checking wind direction doubtless didn't help, either. It wasn't long before not just our factory was completely filled with dense, acrid clouds, but, with the metal cylinders still copiously spewing out their contents, the next door building was also inexorably immersed within a black, murky smog.

As far as exercises went, I still maintain that it was actually a great success. The fire crews squirted gallons of water, the first aid teams applied yards of bandages and countless make-shift splints to our many pre-instructed 'casualties', and the unfortunate 'bodies' were carefully laid out on the far side of the factory yard. Even the workers, eventually crawling out from their shelters to the accompaniment of much loud coughing, and ever-so realistically reddened and streaming noses and eyes, seemed glad to find that they couldn't re-enter either factory for several hours, and thus were presented with a wholly unplanned afternoon off.

Only the management seemed rather ungratefully upset.

All in all, therefore, I had quite a busy - at times, even enjoyable - war. But eventually, and most definitely thankfully, it ended.

That the war was coming to an end had been quite obvious for some time: and Coles, like many other companies, were already being encouraged to prepare for the post-war market. To this end, we started to look for a range of new models, each of us in the office being given a specific and separate job, either singly or in co-operation with another man. I was allocated, and successfully achieved, the sole designer task for a small, one-ton, fully mobile, self-propelled crane, something which we hadn't ever tried before, the smallest model previously produced having been three times the weight. It seemed that I was now carving out a good future for myself at my new, if unintended employer.

Until, only a short time later, I was begged by a very busy Fletchers to return. It was very gratifying to know that I was wanted; and, all in all, it was an offer which I wanted to accept. The problem was, I'd already taken on another commitment. I, along with Andy Lesser and Mike Miller, colleagues met at various times both through school and my current and former firms, had just formed a company to carry out 'technical art services', utilising our wartime skills and experiences to produce drawings, often of a three dimensional cut-away type, and to compile and write illustrated operational manuals, parts lists, and various other kinds of related engineering businesses' literature and leaflets. We'd already acquired office space, right opposite Derby Cathedral, and were spending all our evenings building up our clientele - and, in my case, undertaking all of the company secretarial work, too - until there was sufficient for one of us to transfer there full time.

But Fletchers, to their credit, or desperation, still agreed to let me rejoin them, on the understanding that if and when I was needed in the new venture, I could leave again. And so, in very short order, I was back where I'd started, albeit in a more senior position, designing equipment which my uncle, their cane-milling specialist, was himself now too busy to handle.

Little by little, we three entrepreneurs built up our fledgling business, until first one and then two of the others were fully occupied within it. We then engaged a typist/receptionist, followed by additional artistic staff. Finally, it was time for me to take the plunge.

Once again, I bid Fletchers farewell.

Our business continued growing strongly for the next three years, by which time we had a staff of twelve. In the meantime, Andy, Mike and I had decided to enrol at the Derby School of Art, their Layout and Design of Advertisements course offering some much-needed training in an area with which we would doubtless, one way or another, find ourselves connected. There were about a dozen other students, all younger than us, and all much more concerned with 'mere' drawing, rather than any specific application of the course's content to their actual working lives. We, on the other hand, continually pressed the lecturer with questions on any and every aspect of the subject matter, taking up so much of his time that eventually, much to the others' chagrin, we were effectively receiving private tuition. Inevitably, our monopolistic, if undeniably keen interest was politely but firmly commented upon - several times, in fact - and so, feeling that we'd by now gained what we'd set out to learn, we left the remainder of the course to the true artists.

And then, in 1950, a new Labour government came into power. Quite abruptly, fresh orders just stopped coming in. Simply put, no-one was prepared to commit themselves to any more future financial expenditure in the face of a feared wholesale nationalisation of industry. And so, after six months completing our outstanding contracts, we had to face the sad fact that we'd have to unwind all that we'd set up, gradually releasing our employees until only the founding trio remained.

It was none other than Wally Cockburn, my old and original employer, who then came to our temporary aid. Whilst working at the Royal Crown Derby fine china company, he'd been asked if he could find them a designer for a much-needed upgrade to their hollow-ware production line. It was a task which we quickly, and very gladly, undertook. After many trials - mostly in our own offices, and involving lots and lots of liquid china clay - we drew up our plans, submitted them for acceptance, and soon had the fully-mechanised plant put together at a nearby engineering works, tested, and then finally installed. Reportedly, it increased Royal Crown's output more than ten-fold, enabling them easily to meet the growing post war demand for their cups, saucers, plates, bowls, and multifarious other products. Which was very nice for them, and, for a time, a great - if ultimately meaningless - success for us.

It was Fletchers, for me, to the rescue again. Someone there - I never discovered who - had heard that we were struggling. The next thing that I knew was a telephone call, asking if I'd like to re-join them. Reluctantly, and with very mixed feelings, I did: straight into the thick of a yet another frantic period. Still, at least it was also into yet another more senior position, this time formally replacing my uncle as the senior designer for all of their milling plant. That college art course came in useful, too, as they soon had me producing much of their new advertising literature as well; and, as a side result, deciding that I'd better learn Spanish. At the time, Fletchers were directing many leaflets at the South American market. Their translation work was being carried out by the Spanish Embassy in London, but the resulting texts were, apparently, causing serious hilarity amongst some of their recipients, since pure Castillian Spanish words often had some very different, occasionally quite rude or embarrassing, meanings within the other countries concerned. As it was, I ended up studying that language for over two years, to a level of proficiency which became invaluable when I later worked in various Spanish-speaking parts of the world.

But those were plans which I didn't have then. I'd been promised that, within a few years, I would succeed to the position of drawing office manager. Early in 1955, the post became vacant. But the managing director - as is sadly so often the way of things, then and now - instead decided to bring in one of his cousins, most recently employed in running a Kenyan sugar plant. There was nothing wrong with him as a man - in fact, he was quite a nice person to speak and deal with - but, as these types of things also so often play out, in this particular new managerial role, he was totally incompetent. The loss of morale amongst the drawing office staff, myself included, was huge, its true extent being illustrated one day when a colleague thrust his newspaper in front of me, one finger keenly tapping towards an advertisement. The post of assistant chief draughtsman at the West Indies Sugar Company - a sizeable subsidiary of the giant sugar group, Tate & Lyle - was far above his current capabilities, he stated modestly: but he and others in the office were quite insistent that it was well within mine.

Many domestic deliberations and one short interview later, WISCO's letter of offer, dated 5th of August 1955, was accepted.

The main cause for concern was that my new employer's offices were in London, right in the heart of the City proper. My parents had, of course, already relocated 'down south' (albeit still north of London) during the war: but my sister and my wife's family remained firmly stationed in and around Derby. Then again, we now had travel options of both the train and the motor car: one of which, a fine old workhorse of a Wolseley 14, I now proudly possessed. We decided to take the risk. Estate agents were quickly called, details of suitable properties were soon posted out to us, and brave Ellie was dispatched off alone on a flying - well, railway - reconnaissance mission, as a result of which a house was equally-swiftly purchased on the outskirts of a nice town on the Surrey borders. It wasn't then long before our three teenage daughters were transferred into their new schools, I settled in to my daily train journey from the local station and back, and we all began to adapt to a very new way of life, the whole pace and timing of 'Southern' suburban commuter culture so very different to what it was just a hundred and fifty miles - if still a good day's travel - north.

About three years into my WISCO term, a change in senior directors resulted in one of them insisting on bringing in with him his own design manager. It was, to say the least, somewhat of a snub - again! - not to mention a personal career threat to my then post of deputy chief design engineer. Fortunately, however, I was still kept on in pretty much my existing role, and the new man, Peter Wade, in due course became a close, lifelong friend.

It was Pete, in fact, who, a few years later, recommended that I should take over as the newly-reorganised West Indies Group's overall controller of agricultural, field, harvesting, and transport mechanisation: a new, rapidly-developing, and expanding sphere of interest and activities both for my employer and the sugar cane world as a whole. And thus my working life took another unexpected turn, as I soon found myself repeatedly, sometimes near-constantly, spending weeks to months away from home, travelling firstly around our West Indian operations, then increasingly out to other countries, and other owners' estates.

It wasn't, as it transpired, entirely the best of timings, coinciding as it did with the somewhat-unexpected arrival of our fourth child: something which, to be fair, I don't think that I can blame anyone else apart from myself.

Stemming from that time, and the increasing responsibility being devolved to local estate field and engineering personal, Tate & Lyle began to incorporate my little slice of its empire into a separate specialised international management and consultancy arm, our overall services in time expanding to encompass not just sugar cane, but all agricultural crops to which we could apply our knowledge and skills. Our offices also moved to out to Bromley, Kent, meaning that I now had to drive myself to work each day; each day that I was actually in the UK, at least.

And there, finally, I remained: in terms of base of permanent employment, anyway. As a director and divisional manager of one of Tate & Lyle's technical subsidiaries - if not, by far, the only capacity in which I spent the rest of my working and travelling days - I took charge of all the manpower services which we provided to other sugar estate owning companies and organisations, as well as remaining foremost a specialist consultant in, and retaining overall responsibility for, all world-wide mechanised and agricultural cane-related activities, overseeing both existing and new developments, large and small, for whoever they were being carried out. As an acknowledged world authority - which, I have to admit, still sounds very nice to my ears, even if it was a rather small and select field - I was also increasingly asked to attend all kinds of international conferences, seminars and symposia, giving talks and presenting formal papers to my colleagues and peers on various matters pertinent to my experiences and endeavours; and, ultimately, contributing to the formation and founding of a British sugar industry training and academic body for those yet to follow in my wake.

It could have ended very differently. In the early 1970s, having just returned from a full-time residential year in Brasil, I was offered another similar post, this time heading up T&L's new Miami venture. One way and another, however - not least of which was concern for our UK-based family, both young and not so much - we decided that it wasn't for us. So it was instead Pete Wade who took up the temporary appointment; and chose to continue when it very quickly became permanent. The company in question didn't actually survive the full period to my retirement. Pete, however, along with his wife and children, thoroughly enjoyed their relocation, and decided to stay in the US nevertheless.

Sometime in that same general period, I was also invited to become a director of A. & W. Smith, T&L's machinery manufacturing division. This was another move which, for various reasons, I declined. It would have been quite a satisfactory conclusion to this tale, though, because, shortly after my retirement, Smiths ended up being acquired by one certain Fletcher & Stewart Ltd, fine sugar machine manufacturers of Derby.

2

WAGGLING WINGS

Flying is fun, they used to say. My father wholeheartedly disagreed. He'd sampled the experience once, back when I was still a mere babe in arms: and once was way more than enough.

At that time, in the early 1920s, the only permanent civil airfield in England was located near Croydon, south of London. Up in the Midlands, therefore, we never got to see anything which could remotely be called a passenger aircraft: if such a thing really even existed anyway, the most modern 'airliner' still being a single-prop-engined, plywood-constructed biplane, capable of carrying numbers firmly stuck in the single digits. Occasionally, however, towns and cities around the nation would be visited by a 'flying circus', pressing into service whatever large, open, and reasonably-flat local grassland they could form into a makeshift strip; preferably, set well on the outskirts, and safely away from all habitation. The actual circus attraction, in fact, usually comprised only one, maybe two, twin-seat, wartime-model biplanes: but the atmosphere, the excitement, the anticipation, and the undoubted showmanship still fit the word well enough.

It was Sir Alan Cobham's version which came to Derby in 1921. Twenty draughtsmen in my father's office had each paid a shilling into a sweepstake, raising the money for just one of them to take the ten-minute flight out from the village of Sinfin, over the main city, and hopefully back again. Dad said that he'd only joined in to make up the numbers: but, to his horror, he ended up winning. He did, apparently, try to get out of it, very graciously offering on several occasions to have the lottery redrawn. But his stalwart, gentlemanly colleagues were having none of it, instead all insisting, come the appointed day, on accompanying the lucky adventurer to the very-aptly-named Hell Meadow, where the young, dashing pilot, clad in leather sheepskin jacket, flying helmet and goggles, personally conducted him into the open cockpit, and tightened down the straps.

My mother, still nursing her firstborn, waited anxiously back at home, fervently praying that she wouldn't be widowed quite so early in life.

Dad, needless to say, returned back down to earth perfectly safely: but he never again set foot in an aircraft for the whole of his life. I don't think that he actually saw anything of Derby from the air, either. You can't when your eyes are screwed tight shut in terror.

From my more youthful viewpoint, the sight of anything in the air always both excited and fascinated me, and I yearned to be a part of it. Aged six, in the grounds of Temple Newsam house in Leeds, I saw a huge gas-filled balloon ascending into the skies. Soon afterwards, to my amazement, two men, clearly intrepid daredevils, leapt out of its basket and parachuted back down to the ground. Well, actually, intentionally or otherwise, they landed in the middle of the lake: but I wasn't going to be the one to pass judgement.

Not that many years later, I was on my way home from school, passing through the parklands of Derby Arboretum, when the new airship, the R100, slowly appeared overhead at a surprisingly low height. Its long, fat shape seemed to hover above me for an infinite length of time as I walked along, my head inclined constantly upwards. I had actually already glimpsed its ill-fated sister ship, the R101, whilst visiting my Bedfordshire relatives on a train which passed its Cardington base. Both of these dirigibles, built by different companies, with different designers, and filled with different gasses, had been commissioned by the government to become the new, much-hoped-for forerunners of long distance air travel. But things, of course, didn't quite go to plan. I'd stood so many times amongst the assembled crowds in the Arboretum, watching in rapt attention and waving frantically up at the pilots as their little circus planes skimmed over our heads. How could any of us, back then, have known that it would actually be the future offspring of these tiny, fragile curiosities which would eventually go on to open-up the benefits and opportunities of fast, world-wide travel?

And how could I realistically ever even have thought to dream that, within just a few decades, I would also so easily and routinely be flying up there with them - not just over Derby, but all around the globe - simply as the rather mundane means to get to and from, and around and about, my places of work?

The chance for my own first inaugural flight - indeed, bar one brief day-return sea trip to Northern Ireland, my first ever excursion outside of mainland Britain - finally came in 1954. Fletchers, my then employer, had asked me to travel out to Mauritius, a small island far off the south-eastern African coast, to report back on a problem with one of their mill-driving engines. This wasn't an undertaking without its risks in those days, air travel not least amongst them; a daunting prospect both personally, and for the wife and three young daughters whom I'd have to leave behind. The distance alone was staggering; the country, customs and culture, not to mention food, wholly alien and strange; and all to be undertaken wholly unaccompanied, left to face up alone to whatever this unknown, dangerous new world - not to mention the intervening transportation - might throw at me. Still, very few men - and it would only have been men, in those days - would ever have been presented with such an opportunity just in travel terms alone, never mind the insights and work experience which it undoubtedly offered: the ability to gain some rare, first-hand experience of the raw cane sugar manufacturing process, and the chance, finally, to see in real-life action all of the equipment and machinery which my colleagues and I had designed, and our factory had built. I was determined not to turn the honour down.

One September day, therefore, lugging along my large, heavy, solid-sided suitcase, I took the train 'up' from Derby to central London - in itself still a significant, unusual and exciting journey for many - and then caught a bus for the long ride out west to London Airport, then a mere row of wooden huts set alongside the single-carriageway A4 road. Proper permanent structures were clearly under construction though, and were soon, I was told, to be opened. In the meantime, there was just one very small, and very basic, departure lounge, catering for all passengers, irrespective of their class of travel, airline carrier, or destination. It wasn't very well sound insulated either: and as all of the aircraft there that day were still propeller driven - the first jet liner had only just started to enter commercial service, and wasn't yet at all common - it was, at times, especially when they were throttling up, extremely noisy indeed.

My expedition's initial leg took me to Paris, where I transferred to an Air France Constellation, a four-engined, cabin-pressurised plane which held about sixty passengers. Air Chance, as many called it, ran the only route to Mauritius at the time: and they took, by today's standards, a very long time to do it. On the other hand, the whole flying experience was much more leisurely. No food was served on the plane, bar some small snacks and drinks. At each mealtime, therefore, the plane had to land - presumably it, too, by then needed feeding - and we had to be taken to a suitable venue. Thus we stopped firstly for lunch in Rome Airport's very fine restaurant, went by coach to a suitably-upmarket Cairo hotel for our evening dinner, then flew on through the night to Nairobi, where coaches again took us to an equally 'posh' country club for breakfast. I can't now recall where our second mid-day meal was provided, but eventually we braved the wide ocean crossing, and reached our final destination just in time for a rather late dinner at our respective hotels, followed by a much needed, properly-lying-down sleep in a real, flat bed.

That latter omission aside, the 'Connie', as a means of transport, was actually very comfortable, and the whole flying experience, multiple take-offs and landings notwithstanding, gave me no worries whatsoever: excepting that, during our initial daytime flight, I noticed several dark streaks appearing along the top surfaces of the silver wings, stretching back from one of the engines. "Oil," the engineer in me immediately thought, followed swiftly by: "Should it really be like that?" Apparently, it presented no particular concerns to the crew, although I did notice that, after returning to the aircraft after each refreshment stop, someone had wiped the wing a little cleaner again; and, I hoped, also topped up the oil.

That first night's journey was, to be fair, a little more disconcerting. The plane's radial engines had short, open exhaust pipes, venting both above and below the wings. It was the former which presented me with the most considerable worries, as, all through the long, dark hours, tongues of flame could be seen spewing out at a point just above the wings' upper surface. Still, I reasoned - and definitely hoped, and probably prayed - this type of aircraft, and these models of engines, had been operating for a long time now, and the exhausts hadn't yet burned a hole in the wings. It should, so I rationalised, be safe enough. Over the top of something which was by now also covered in a liberal coating of oil...

It was exactly the same coming home three weeks later: oil, flame, stop-overs and all. Quite probably, it was the same, even maybe the airline's only, plane. But by then I was quite the experienced air traveller. Wasn't I? (Note to readers: it turns out that both oil leaks and flames were well-known features of the Connies, and appropriate steps had indeed been taken to put additional fire-proof plating in the appropriate places to prevent the otherwise rather alarming side-effects of their engines' exhaust gas recycling, turbo-charged design.)

It was New Year's Eve, 1963, before I took another flight: this time to British Honduras (now Belize), and thence onwards for a whistlestop familiarisation tour of various Jamaican and Trinidadian estates. London Airport, then also conveniently closer to my new home, now possessed three entirely separate, and significantly-larger, terminal buildings. Many more types and numbers of planes were flying in and out, too, the biggest being the Boeing 707 jet, one of which I duly boarded. This time, we didn't have to land at all until Washington DC, not only to refuel, but to attend to all of the necessary US immigration and customs clearance procedures - which in themselves probably took about as long as one leg of the earlier Air France trip - before flying straight down to New Orleans. And there, for the first time, I met one of the less attractive aspects of this mode of modern travel, for we must have spent almost as much time circling and waiting to land as we had in crossing the Atlantic, a spell of unexpected wintry weather only compounding the inevitable backlog of too many planes all arriving at the same time, and the standard confused, last-minute US travel rush of everyone desperately trying to get to wherever they wanted to be for the imminent festivities.

By 1966, the still-expanding airport had been renamed Heathrow, and it was quickly gaining a competitor further south in the Surrey countryside. Gatwick's fast-growing facilities now offered the international traveller an increasing choice of carriers and routes to suit all needs, timescales, and purses; well, in theory. My scale and sphere of operations had also grown, and my overseas travels, starting with an initial extensive tour of the islands of the West Indies, the southern USA, and then Zambia, began in earnest. So much so, in fact, that all those long transoceanic, transcontinental journeys in my stalwart, trusty and reliable friend, the Boeing 707, soon became wholly routine, and mostly uneventful.

I was, from the start, always a very organised traveller, and, fortunately, also a good onboard sleeper, finding it easy to adjust to the problems of jet lag. One way and another, therefore, I very quickly became rather blasé about the whole business. (As long, of course, as you didn't mention Christmas, a word which, in my personal opinion, should never, ever, be used anywhere near the word 'flying', except when the words 'not' or 'never' are also firmly in place.) I did, however, re-experience just a moment of that youthful excitement when catching my very first sight of the gargantuan, how-can-it-possibly-stay-in-the-air, Boeing 747 'Jumbo' - one of the first in service, I believe - as it lumbered in to land at Puerto Rico. Compared to some of today's giants, of course, it was but a mere baby.

It would be good - or, at least, the expected thing - to comment at this juncture upon just how much things have moved on and improved since my earliest days of air travel. In some ways, of course, they have; markedly so. Safety standards, navigational and communications systems, developments in electronics and control technologies, automatic landing systems, airport and runway layout and design... Yes, I'm very thankful for them all. And yet, in other ways, nothing has really changed at all; or, at least, it's all come around again. Whether whilst writing this, or whenever you get around to reading it, the news somewhere or other, for some multitudinously-varied reason or other, is all too often full of horror stories about delays and last minute flight cancellations, stranded passengers, ground and cabin and cockpit crew strikes, the rising costs and shortages of fuels, lost and piling-up mounds of un-handled baggage, snaking hours-long queues both to check in, to get through airport security, and then, upon one's return, the varying delights of trying to locate and reclaim your luggage; and so on and so on and such forth. Then there'll be requirements and exclusions for nationals of certain countries, requirements or exclusions for nationals of every country, variants of different kinds of requirements or potential exclusions for nationals of or entry into - never forgetting exit out of - various different countries, all kinds of general and doubtless different-for-every-single-country requirements for vaccinations or proof of funds or required minimum or maximum lengths of stays or purposes of visits, never mind your standard health scares and disease outbreaks, or whatever quasi-religious or political disturbance or falling-out happens to have happened at any given place or moment... La plus, I suppose, can really ever change.

It is, however, true that my days in the air, even those near to my retirement, were certainty nowhere near representative of the experience with which most current-day readers would be familiar, whether for pleasure or business. There were no automated or electronic check-ins, never mind online systems - there was, of course, no 'online' at all - and no 'budget' airlines and whatever ingenious schemes and practices they continually think up to make your travelling experiences all the more exciting. Airport security, certainly during my earlier travels, varied from the non-existent to the pretty basic, and was usually fully manual: the larger airports were latterly equipped with X-ray machines, but there certainly weren't the rafts of bleeping gates and body scanners which are commonplace now. All too often, your onboard safety was assured merely by a man having a good rifle through your carry-on baggage - and that, half the time, was more to ensure that you weren't smuggling out something appropriately illicit, or carrying more cash than permitted - occasionally asking a few annoying questions - OK, so maybe that's not changed so much, either - and occasionally patting you down with some degree of thoroughness before wishing you a pleasant flight. Most airlines, if not necessarily airports, did generally draw the line at carrying larger knives and firearms; on your actual person, anyway, although handing them to a stewardess when boarding was usually considered sufficient to ensure everyone's safety.

Once within the aircraft, even back in second (but most definitely not economy) class, one was always treated as the proud and valued passenger which, of course, you were; on the larger national carriers, at least. You didn't need, and sometimes weren't allowed, to take much into the cabin with you, except for a coat or jacket, a good thick reading book, your valuable and delicate binoculars and camera, your passport and other essential papers, and whatever overnight toiletries you might require: were they not in any case provided as a matter of routine courtesy, the passenger finding awaiting upon their seat a good-quality, logo-embossed washbag containing, at the minimum, a toothbrush and small tube of paste, a comb, small bars of soap, a hygienically-bagged facecloth, and perhaps even small towel and some cologne. Plentiful free food and drink was soon brought around for one's comfort, main meals were served only once you'd selected your choice of starters, mains, desserts, and wines from the oft-illustrated menu card, and, once at cruising altitude, a general invite was usually put out over the cabin's tannoy system for anyone interested to visit the flight deck and chat with the pilots: an experience particularly enjoyed by the few youngsters who'd occasionally be seen flying, as well as both the novice and regular adult passengers. Later in my travels, there were even large cinema screens which pulled down across the aisle, and upon which one could watch - free of charge, if not choice - the latest cinema releases, listening via a set of headphones which the stewardess would hand out, and only occasionally collect back. Pillows and blankets would be provided for those wishing to recline and stretch out on overnight flights - and the seats and legroom were most definitely sufficient to allow you to do so in relative comfort, notwithstanding the fact that the planes often flew far from full, and therefore offered the lucky traveller an entire row over which to sprawl out - whilst complimentary satin eye shades were usually found within the comfort packs. I never remember ear plugs, though: but, to be fair, that was also generally before the days of too many screaming babies and children, rowdy holiday makers, and the invention of any kind of those infernal music-through-earphones machines. Hot scented towels would be brought around the cabin for those wishing to refresh themselves after waking, and, of course, after meals: an especially-useful aid if you'd been using the beautifully-engraved, and also never re-collected, chopsticks provided by most of the Eastern airlines, along with their ornately-patterned silk fans, with which to waft around the hot air in the absence of overhead blowers. (I still have many of them, along with an older Air France cardboard one, too.)

Even the 'carry-on' luggage might be provided, my trusty dark blue 'BOAC' shoulder bag making almost as many trips as I did, still transporting my many washbags, headphones, and assorted eyeshades long after they were all sadly discontinued, as air travel became more and more a simple commercial transaction, and such costly frivolities were reserved only for those prepared to pay dear for the privilege. (Said faithful bag, by the way, upon its eventual enforced retirement, was replaced by a very fetching small, and entirely-useless except for passport- and paper-carrying purposes, leather zip-up 'man bag', as my wife instantly named it. These, carried either over the shoulder by means of a thin strap, or more usually dangling from the hand by means of a similar wrist attachment, were very popular for a time with some sectors of the male travelling public, avoiding all that annoying spoiling of the cut of one's jacket by the shoving of necessary documentation into an internal pocket. I was duly presented with such an accessory as a gift from one of my son-in-laws, quite late in my career. Frankly, I found the thing rather embarrassing. Useful enough, I'll admit: but it still travelled firmly tucked inside my jacket.)

The whole general process of transporting one's possessions had, of course, moved on considerably from the days of my childhood's family holidays, when, usually about a week in advance of our scheduled train journey, a uniformed railway company porter would arrive at the front doorstep, along with his horse and cart, to collect the pre-booked number of extremely large, and very heavy trunks which contained all of the family's required clothing - often pretty much all of the clothing which our entire household possessed - and other essential items for the planned stay. Still, at least we could be confident that they would always be found awaiting our weary and smoke-begrimed arrival at our chosen boarding house; and also that, upon the return leg, they would be collected on the day of our departure and reliably returned home in good order, undamaged, the following week.

Which is more that I can say about a fair number of those oh-so-much-more-efficiently-transported cases of my flying days. To be fair, given the frequency with which I travelled, never mind to and via where, or whatever number of oft-tight connections which had to be made in-between, I don't remember actually ever having truly lost anything, nor even having had anything badly damaged, never mind stolen: and maybe that's a minor miracle in itself. But there were certainly a good number of annoying delays; and several near-misses. Two days without baggage was about par for the course for most airlines. Pan Am, as well as taking the prize for the biggest repeat offender, held the record at ten: although they did, in the meantime, reward me with the then moderately-generous sum of US$30 for my troubles. This, at the time, was enough to buy not only sufficient replacement clothing, but also a suitcase to put it in, such that I could still fly onwards from Puerto Rico on what was only the beginning of a very long, and extremely-rushed trans-Caribbean tour, my subsequent itinerary so convoluted that I did at least have to give them credit for managing to keep enough track of my movements ever to return my luggage at all. It wasn't the first, nor last time that I'd had to start my field studies by wading through the scrub, bush or cane fields wearing whatever I'd travelled out in from the UK: and if those clothes just happened to have been designed for a full English winter, whereas it was the height of summer in the hot, humid tropics... Well, at least the local staff, attired only in thin cotton shorts and shorts, had a good laugh at my expense.

I quickly learned to pack duplicates and back-ups wherever I could.

I also learnt the art of double-checking destination luggage tags: not only at the check-in desk, but at the several other times when I'd often again have the chance to assess my belongings' welfare. I'd frequently end up changing the tags over, too. Lagos, Nigeria, was a particularly good case in point; a whole trolley-full of cases, actually, the assembled, assorted, already-checked-in hold luggage being physically passed by every time that one entered the departure lounge. I do stress assorted, however, because, in that strange way in which the Nigerians operated at the time, they had decided that it was most efficient to load every single passenger's luggage onto the same train of trolleys, irrespective of airline or destination, before taking it out for loading into the various awaiting planes. But at least it gave you the opportunity to locate your bags, check the labels, match them with those of your travelling colleagues, and effect whatever other hands-on interventions were required to ensure that all sets of baggage were similarly tagged, all were placed right next to each other, and, therefore, all stood their best chance of going forwards to the same, hopefully-correct location.

Yet even in New York, transferring from one airport to another by the, for me, novel means of a helicopter, with my bags already booked and labelled all the way through from my original boarding point to my final destination, I still had to intervene when spying my suitcase in a pile awaiting collection at the wrong exit gate. This time, however, it wasn't such a simple DIY switch - for which security measures I really ought to be thankful - the pilot and other passengers on my onwards transport having to wait ever-so patiently whilst I tracked down a ground steward and arranged for the case to be put on board along with its owner.

The Thais also had a particular and peculiar trick which they like to play on unsuspecting foreign travellers wishing to depart their country. Having checked-in at the desk and obtained your boarding card, one then proceeded through the usual emigration and finance checks; and then, suddenly, once again found yourself being presented with your hold luggage, only recently handed over, waved off into the airline's trusty care, and by now, presumably, already well on its way out to your plane. But no: for here, in a type of reverse customs arrangement, everyone was at this point required to open up their cases for examination. Only then could they finally be loaded onto the waiting luggage trolleys. This, apparently, was an attempt to prevent removal of the country's antiquities and heritage items: but why it couldn't take place at or before check-in, I never worked out.

Of course, especially in my earlier forays abroad, many 'international' airports were still little but wartime Nissan huts, or at best small concrete or metal sheds. Shiny, or at least decently-solid, newly-constructed terminal buildings were, however, starting to spring up quickly; the supporting infrastructure and ground services, on the other hand, was usually left lagging somewhat behind, an obviously-clear afterthought, if not overlooked completely. It was therefore quite common to have to collect your bags from the tarmac beside your aircraft, if not sometimes actually load and unload them yourself, and then carry them into the customs area to be inspected. Then again, there was a good deal to recommend it in terms of speedy baggage reclaim.

The Portuguese airport of Oporto was also keen on efficient off-loading, in the process creating the most wonderful chaos all of its own. It was, to give them their dues, an unintended event for all concerned, the transatlantic leg of my pre-Christmas trip back to the UK having been diverted from a fog-bound Lisbon; as also were three other, similarly-large airliners, all landing within minutes of each other, chock full of tired and already-frustrated travellers and their many and multifarious possessions. It was only a local airfield, and they weren't used to handling anything of this capacity at all: but they certainly gave it a darned good go, simultaneously managing to unload every single bag from each plane onto their solitary, small, circular conveyor. The assembled luggage, a random mix of all kinds of departure points and onwards destinations, then merrily travelled once, and only once, around the primitive, and soon heavily-overloaded carousel - itself now surrounded by ten-deep crowds of confused, stressed, and generally-uninformed passengers - before simply cascading off at its end and skidding outwards across the slippery, polished-marble floor. The resultant swiftly-mounting piles very quickly became so large and precarious that it was too dangerous even to try and step or reach in to recover one's cases - also usually both very heavy and hard-sided - without others falling straight on top of your head. No-one thought to, or perhaps for some reason couldn't, stop the conveyor, so it was only after every single item had been delivered and deposited that the bags could finally, one by one, very carefully, be tugged out, their tags examined, mis-pronounced names on labels called out, and eventual reunification with their owners achieved.

Taking off and landing are said to be the most dangerous times of any flight. For me, though, changing planes, and usually airlines, often provided the most challenging, amusing, and, occasionally, traumatic problems; not only, by far, as concerned my luggage. Back then, there were precious few direct flights between anywhere, so catching an onwards connection at some random transfer point was often required, even for a fairly simple journey between two large capital cities.

The most frequent issue was simply the failure of at least one plane to fly in or out at the appointed time, the result at best being the receipt of some very dubious directions to some secret access or emergency door, an all-out sprint across an unknown airport, supported by encouraging shouts and occasionally-helpful directions from the ground staff, and, if lucky, one's almost literal hauling, forcible seating and buckling into the waiting aircraft as it began taxi-ing out. It put a whole new meaning on the phrase 'fit to fly'! The luggage, needless to say, usually followed on later.

In Lagos, of course, the full DIY option was again in favour. Once, a colleague and I became unexpectedly held up for nearly an hour, queuing for the compulsory indignities of a currently-in-vogue full body search. At last released into the departure lounge, we were further delighted to discover that our flight, a local 'short hop' service, had already been boarded, and could now, in fact, be seen rolling out towards the runway. And so, with no-one and nothing to stop us, we both tore out of the 'gate' door after it, and sprinted straight into its path, frantically waving our arms to flag it down. As, of course, you do. Fortunately, after a couple of visible prods and points from within the cockpit, the plane came to a swift halt, the door was opened, the ladder let down, and we were welcomed aboard and shown to our empty seats by a smiling, and thoroughly-unperturbed stewardess. They'd thought that they might have been two short of the checked-in passenger manifest, she and the pilot shrugged: but it didn't really matter either way. That time, of course, our bags were already on board.

Transiting the USA was a rather more formal, not to say fraught and complex affair. Travelling out to Belize one day, I duly 'deplaned' at Washington DC, collected my cases, cleared through immigration and customs, handed the bags back to the airline, and then began the long walk through various parts of the sizeable airport in order, eventually, to await re-boarding the exact same aircraft again. The USA hadn't yet updated themselves to the procedures by which the rest of the world was by then operating, and didn't provide a dedicated transit lounge for passengers not actually intending or needing to enter the country itself. Perhaps, we often wondered, that was the entire problem: as the USA was clearly the centre of the then-known universe, they just naturally assumed that everyone must have wanted to go into it, instead of using it merely as a staging post to go on elsewhere.

On this particular occasion, they didn't seem to want to let me go anywhere else at all, for, upon entering the corridor to the departure lounge, a large man dressed in dark civilian clothes stepped purposefully into my path, produced some kind of security services ID card, and enquired in a rather demanding and far-from friendly manner as to who I was, where I was going, and why. I answered his questions easily and truthfully: but he didn't seem satisfied. He then asked - rather strangely, I thought, for a trans-Atlantic traveller inside a large international airport - if I could prove who I was: so I produced my trusty battered, tattered, heavily-thumbed, and many-times-stamped British passport. (Well, one of them, anyway. Duplication, sometimes triplication, was often necessitated by the many and various times when country A wouldn't allow a citizen of country B to enter if they'd ever set foot in, or carried a passport bearing an entry stamp or visa from, country C, or some similar variant upon such a general geopolitical theme, dependent upon who exactly wasn't talking to or approving of whom at any given moment.)

"Anyone can fake one of those," my inquisitor then stated, in superciliously-suspicious tones.

So then I showed him my business cards, only to have them provoke exactly the same response.

"Anything else?" he enquired, mockingly.

I was seriously beginning to wonder what was going on, who or what they thought I was, and why exactly they were so worried about it. In desperation, I searched deep into my briefcase and found letters addressed to me by our clients in both Belize and the US, together with several commercial and published research papers within which my name was mentioned.

Eventually, and evidently very unhappily, he did let me go on my way: although he quite unsurreptitiously followed me all the way to the departure gate, and didn't turn his gaze away until I'd physically boarded the plane. Whether he then had me watched when I subsequently passed through New Orleans and onto my connecting flight I'll never know: but it was definitely an un-nerving experience, and left a rather nasty taste in my mouth.

This wasn't the only case of mistaken identity which befell me: although, in the other case, it did have its distinct advantages. To some observers, I apparently bore a passing resemblance to a certain senior British politician-cum-diplomat of the day. From my point of view, he just happened to share a very good taste in his choice of fashionably-respectable eyewear. Either way, more than once, mostly within the US, I found myself being hustled and ushered with all due and time-saving VIP and security privileges all the way through to my onwards flight; constantly, of course, protesting that I was not the said noble and honourable gentleman in question, but a mere fellow countryman on private company business. They were certainly the fastest, most pleasant customs and security clearances which I've ever done. And maybe, as regards Washington, just the odd bit of payback.

Assuming that you did actually eventually manage to arrive at your intended destination - which I always did, if sometimes a little later or less well-attired and equipped than intended; and only once with everything reeking of rum - the seasoned commercial traveller's next obstacle was to run the gamut of the many and infinitely-varying immigration, inwards customs, health, and random other border checks and procedures. Normally, although always acting through our UK employer, my colleagues and I would be carrying out our work at the request of, or directly on behalf of, some kind of government ministry - assuming that they hadn't had yet another coup whilst we were on our way - or other large national association or corporation. In general, therefore, as long as our visas, inoculations, appropriate amounts and types of currency, proof of accommodation, introductory letters, and all sorts of other requisite documentation were in order, there shouldn't have been any issues standing in the way of our admittance to the country concerned. Officialdom, however, no matter where one is, always seems to have its own ideas. Whether it was misunderstandings, incompetence, cultural and language differences, an unfamiliarity with various aspects of everyday Western business or monetary practices, a certain in-country political difference of opinions with the current ruling party, or just outright and often overt corruption, they each had their own peculiarities; most, generally speaking, just to add to the overall fun, given to changing without too much notice upon some random legal, administrative, military, or other, perhaps merely personal, whim of the season, week or day.

But even once all of the appropriate evidence had been produced, the multitude of questions had all been satisfactorily answered - with or without the less-than-occasional raised and quizzical eyebrow - and I had finally been granted permission to enter, that still wasn't the end of my travels by far; nor of their problems. It should, surely, have been easy enough. All that I had to do was to present myself at my initial, pre-assigned, pre-arranged place of work - sometimes a city office, sometimes a large factory, sometimes straight out to some remote site miles from even the most basic civilisation - or, perhaps, interim overnight accommodation. The theory, so I was repeatedly assured, was that everything would already have been taken care of well in advance of my flight's arrival, and that I would be met at the airport by the appropriate company's or ministry's local representative, an in-country colleague, or at least a car and driver, and taken directly to wherever I was to stay. The practice was that I became quite an expert at standing and sitting around airport lobbies, from the dusty, makeshift and ramshackle to the most modern and newly-constructed, rarely knowing even where I was meant to be going, never mind whom to contact or how to get there, armed with nothing but hope and a very thick novel.

Getting back to the airport frequently wasn't without its travails, either; nor the surely-simple process of getting out of the country again. That, in fact, often posed a lot more problems, and created a lot more trouble, than the getting-in, if pretty much for the same reasons. In some countries, it had become developed into an entire artform all of itself.

Thinking back on it, maybe that last paragraph's opening assumption wasn't altogether wide of the mark after all. Each airline, in my experience, had its own character, its own marketing and branding, its selling points, its own status and image; or, in the case of a lot of the smaller overseas ones, just downright distinctly 'interesting' practices, and a certain reputation all of its own.

Air Afrique became particularly well-known to me, as did the requirements of their general embarkation procedure, in turn linked to their established habit of always selling several more tickets than there were seats available. Upon being informed that the flight was ready for boarding, therefore, the hopeful passenger, even at main international city airports, was most strongly advised to make a literal headlong dash, alongside everyone else also wanting to travel, out across the tarmac, up the steps, and onto the plane. Rather like some children's party game, the unlucky or unfit ones - or those having not already simply pre-bribed the on-board stewards - without at least a firm hand, if not several pieces of luggage or clothing, immediately secured upon the nearest unoccupied seat, were simply turfed back off and left behind; even if their hold baggage was already loaded.

According to the admittedly-perhaps-apocryphal story, the first Zairian pilot to fly an Air Afrique jet into Paris was told by French air traffic control - as was, of course, quite standard in their crowded skies - to go into a holding pattern and await a landing slot. But allegedly, so it goes, the said pilot declined to comply, instead proceeding straight down onto the runway. When later questioned, he replied simply that he was never going to be told to wait behind any white man, no matter what the reason. Thankfully, I don't think that this particular airline policy, at least, ever caught on.

Aeroplanes are, of course, expensive things. Consequently, some were, to put it mildly, not in the most confidence-inspiring of conditions. But other carriers came up with better solutions. On my first flight across from Johannesburg into Mozambique, for instance, I boarded a plane decked out in the livery of DETA, the local Mozambique airline. We stopped on the way at Swaziland's Manzini airport, where we all had to disembark and pass through a transit lounge. It was whilst standing there, waiting to reboard, that I couldn't help but to start scratching my head, for the plane, having already turned through 180 degrees in preparation for take-off, was now, following the quickest paint job in history, proudly displaying the lettering, colours, and insignia of South African Airways. It transpires that this was quite common practice, with both airlines sharing not only the route, but the aircraft itself, a side each.

Nigeria Airways also weren't above sharing, or in this case borrowing, either; nor were they exactly fussy about the type of plane. After several nights of aborted departure attempts on one of their 707s - and the less said about that particular airline's mechanical issues, the better - my fellow passengers and I eventually found ourselves leaving London on board Aer Lingus. It seemed that the Irish airline helped to manage the Nigerian carrier in the UK, and with the latter still unable to overcome their own craft's ongoing issues, the former had agreed to bring in one of theirs. The only problem was, the Irish didn't actually have a spare passenger jet. But, in the circumstances, in either a wonderful or terrifyingly-worrying mix of these two countries' logics, both airlines had decided that a cargo plane would do just fine instead, very hurriedly fitted-out with enough seats, if not yet quite enough seat belts. Still, shrugged the Nigerian crew, it wasn't too much of a worry. If the passengers would just kindly hold on to something as best as they could, we'll all finally be on our way.

It was on that same flight, most of it unusually now in daylight, when, some time after crossing the Mediterranean, the large Nigerian man seated next to me suddenly rushed over to the other side of the plane and, along with several of his fellow countrymen, began peering and pointing excitedly through the window. It was quite a while later when he finally sat back down again, now with a markedly despondent air. Innocently, I enquired what he'd been looking for.

"The Tropic of Cancer," he sighed dejectedly. "It's a line, you know, which runs right around the world. I've seen it in my school books." He seemed utterly and earnestly sincere. "But we must be too high to see it from here."

To be fair, he probably wasn't the only one at that time, or since, to hold such a belief. My own, once-thought-wholly-adequate geography books, after all, had certainly never contained a helpful asterisk and footnote to the effect that the lines of the tropics and the equator, like those of latitude and longitude, weren't actually physically painted upon the earth's surface. For navigational assistance, I've thought so many times since, it might well be a very good idea if they actually were.

Smoking was, of course, still freely permitted on any airline, although passengers were always requested to extinguish all smoking materials during take-off and landing. Just in case... It wasn't a rule which Philippine Airlines seemed to care much about, though; passengers and crew alike. Then again, whilst they operated a fleet of quite modern, medium-sized jets, they did have somewhat of a, shall we say, all-encompassing, slap-happy reputation. Check-in, particularly for their local and inter-island flights, was a very easy-going affair, where they hardly ever bothered to check the tickets, and certainly never weighed the bags. Once on board, prior to a take-off which didn't usually inspire one with any more confidence than their hopefully-subsequent landings, they did usually remember to give at least a nod to the general concept of a pre-flight safety demonstration, albeit always ending with the additional polite request: "And will you please now kindly park your firearms with the stewardess", after which the lovely lady dutifully proceeded along the plane, woven basket in hand, collecting a good selection of the natives' beloved revolvers.

The Indonesians were very keen smokers, too. Only they didn't favour the type of cigarettes enjoyed by most other people; nor even those rather more fragrant and relaxing varieties frequently encountered in some parts of the Caribbean. Instead, by the plane-full, they preferred something called 'kreteks', which, beside tobacco, contained some kind of cloves or other small seeds, adding a certain aromatic flavour, but also rather prone to exploding when ignited, spitting out little flying sparks in all directions. Hastily banned by all major airlines, and despite several mid-air fires, some with quite disastrous results, anyone travelling upon any of that particular country's internal services still had a reasonable expectation of arriving with at least some part of their clothing full of small, singed holes.

It was Air Panama, however, which probably brought me as close to disaster as I've ever been. Along with a colleague, I was flying back to Panama City from a small regional - for which read very rural - airport, its single strip set amidst a low, flat plain of scrubby bush. Once, it might just have sufficed for the occasional odd bit of light traffic: but now, even in the best of weather conditions, the short, roughly-gravelled runway couldn't be described as anywhere near approaching ideal, even before adding in the fact that it started off at the top of a slope, ran downwards into a shallow valley, and then ascended out again towards the other end. Still, undeterred, if not entirely inspired with confidence, we and our few fellow passengers began to make our way out across to the awaiting aircraft, a twin-prop, high-wing, twelve-seater Beechcraft. We left the terminal walking. By halfway out, we were running through a typical sudden, torrential, Central American downpour. The flight, however, was already delayed, and so, with all having scrambled, somewhat sodden, into their seats, and in spite of the white sheets of rain rather more than reducing his visibility, our determined captain immediately began to rev up the engines. Soon enough, and at least in a reasonably-straight line, we were tearing off - or perhaps bumping and bouncing is a better description - literally down the river-cum-runway. Until, of course, we reached its lowest part, where the plane proceeded to plough through - at this stage, just the word 'through' was a relief in itself - what was by now a considerable depth of water, immediately scrubbing off pretty much all of our previous forward speed. Undeterred, with the screaming engines pushed to their stops, the pilot once again accelerated onwards and upwards; upwards, that is, with the wheels very firmly stuck to the ground. By now, of course, I, all of the other passengers, and probably all of the crew, were starting to get a peculiarly-nasty feeling that we weren't going to have a hope of attaining sufficient V-whatever-it's-called to take off before reaching the end of the runway. But, right at its very extremities, the pilot finally managed to haul up the plane's nose just enough that it cleared the nearest low trees by a proverbial whisker. We all swore that we'd heard branches scraping the underbelly, and were very surprised to find that, upon our otherwise-uneventful arrival at the other end, both wheels were still fully attached.

My amusements and concerns weren't, by far, just confined to the commercial airlines and airports. Out of the necessity of our isolation, never mind the need to travel significant in-country distances, or to oversee work upon huge agricultural estates and cane fields, we often obtained the use of our own two-, four- or six-seater plane, and operated our own private airstrip. This, for long-term use, was usually effected through arrangements with the appropriate factory's or estate's management, but sometimes, for the more unexpected or urgent one-off trip, we merely hired one. It wasn't luxury private travel by any means: but it was always - nearly always - enjoyable, the frequent chance to sit up front, right in the cockpit alongside my pilot-cum-everyday colleague, taking me straight back to those early boyhood dreams. Regretfully, I never managed to get around to getting a pilot's licence myself. I always wanted to. I often intended to. I simply never seemed to find the time.

I did learn quite a lot from being ferried around in these smaller planes, though. Firstly, and most importantly, especially when travelling over vast, endless plains, scrub deserts, and swathes of dense tropical rainforest jungles, that it is not only most advisable to check that the pilot actually does know where he's going - a general familiarity with local runway locations, not to mention sizes and shapes, being here an essential inclusion - but also to ensure that he has on board at least some kind of rudimentary navigational aid, such as a map and compass.

This last, ever-so helpful fact was most ably demonstrated when flying over the Amazon jungle. Along with a Brasilian business partner, I'd gone up to the area to look at a prospective new estate site, taking a scheduled internal jet up to the major north-eastern port of Belem, then transferring into an old, but still serviceably-looking Dakota for the onwards, hour-long journey west. The plane was surprisingly full, its twenty or so passengers all accommodated in some kind of intriguingly-slung hammock-type seats; probably a military relic, I thought. All seemed otherwise well until, on the verge of commencing take-off, the engines were brought to an abrupt stop. Returning to the terminal, and after a lot of frantic and angry jabbering, my colleague informed me that the airline were, most annoyingly, refusing to fly this particular plane today. They had just discovered, during their rather delayed pre-flight checks, that their gyro compass was broken, and they weren't allowed to fly without one: a wise precautionary regulation, I was soon to agree. My ever-eager companion, however, being a man of some considerable influence, never mind determination, wasn't going to let such petty problems stand in the way of a major business opportunity. So it wasn't long before we and three others - I knew not then or now who or why - were being ushered out to a six-seater Cessna, already warmed up and waiting, and were soon once more on our way.

The view, even at a fairly low altitude, was nowhere near as enthralling as I'd always hoped. After a certain initial enchantment, the endless carpet of unbroken green, stretching the full width from horizon to far horizon, with not a vestigial glimpse of ground to be seen, quickly became, well, rather boring. Our pilot was clearly of the same opinion, I noticed, beginning to make a series of firstly right, and then left turns onto reciprocal headings. Soon, however, I began to sense that all was not fully well. A certain tension quickly began to manifest amongst the other passengers, too, along with a dawning realisation that our pilot had, quite literally, lost his bearings, and was now adopting some kind of rudimentary search pattern, resorting to quartering the featureless canopy below in the hope of some visual aids. Staring helplessly out of my little window, I started to think more and more vividly about all the stories which I'd heard of planes going unaccountably missing in this dense, disorientating forest, the wreckage often taking years to locate, if at all, the bones of their passengers and crews rarely found. I wondered how long it would be before we, too, ran out of fuel and had to ditch amongst the trees: an outcome which seemed increasingly likely as each endless droning minute passed.

And then, with a little triumphant shout coming from someone - probably the pilot - we saw in the distance the Amazon itself; for all of its greatness, remarkably thin-looking and hard to make out. A shared feeling of relief swept over the tiny cabin; doubtless also the cockpit. Locking onto, and then following the river upstream for about an hour, a left turn brought us parallel to another, still-sizeable waterway, besides which soon appeared a small town. Its newly-constructed airstrip seemed terribly short, sat remarkably close to the edge of the riverbank, and was surfaced only with very bumpy grass: but I've never been so glad as I was when we landed upon it.

Knowledge of the likely intentions of cross-border insurgents, never mind the impromptu plans of discontent, heavily-armed local workers, was also pretty handy; as were the latest positions of all nearby surface-to-air missile batteries. Our frequent 'in-house' aerial surveillance trips whilst on the Swaziland project, for instance, necessitated a route very close to, and parallel with, the border of Mozambique, which wasn't always the most peaceful of nations at the time. These flights, affording us a quick and easy overview of the developing estate's vast fields and general infrastructure, quickly became routine, lazy affairs. Until, that was, our usual smooth course was rudely disturbed by the pilot, one of our own site engineers, suddenly and without warning banking the plane steeply over, and veering sharply sideways.

"Ah," he noted casually, as he straightened up again. "See over there?" Without any particular signs of either surprise or alarm, he pointed down and behind us, towards what was, presumably, indistinguishable amongst the variously green, yellow, and light brown lands below us, another country. "They've brought in some more SAM launchers. They'll be tracking us, sure enough, and if we accidentally fly even a yard into their territory, they'll not hesitate to fire one up." He made a quick note on a pad by his controls. "It'll not be the first time."

I thought back to that Nigerian passenger from my earlier travels, looking down for his painted line. Perhaps in this case, too, a bit of helpful border marking might not have gone amiss.

Still, suffice it to say that from wherever, and for whatever reasons, came my most memorable flying moments, however many odd and interesting things didn't quite go to schedule, whatever risk or adventure I inadvertently ended up undertaking, I luckily survived to tell their tales.

And any landing which you walk away from, as the old adage goes...

3

BELIZEAN BEGINNINGS

New Orleans airport, 31st December 1963. The terminal concourse was jammed with frustrated travellers, the road in-between its outer doors and my hotel so heavily congested with all manner of skidding, spinning, and just plain stuck vehicles that it took me a full three quarters of an hour to traverse its two hundred yards' width.

Given the date, a ten-inch covering of snow might have been considered rather seasonal: but not to the Orleanians, who hadn't experienced the white stuff for many, many years - if, in fact, ever in their lives - and were understandably not at all equipped to cope.

Returning a little more easily to the airport very early the next morning (or, in this case, year), the scheduled 6am service - provided courtesy of a rather rickety, and distinctly antiquated, two-engined craft - took me straight down to Stanley Field, a tiny, rudimentary, and most definitely aptly-named 'airport' cut out from amidst the low-lying bush near to Belize City, capital of the country then known as British Honduras.

It was, at least, much warmer. Having picked up my bags from the plane-side trolley, and quickly cleared the screened-off trestle tables of customs and immigration, I wandered out into the compact concrete terminal building's foyer. Its few facilities were all shut up, the entire place - not that it was ever exactly bustling and busy, even many years later - positively barren and deserted. The handful of other passengers who'd flown down with me soon disappeared off to their own devices, and there I was, the newly-inaugurated international business traveller, left utterly alone. No-one had come to meet and greet me. No-one was there to tend to my onwards needs. No-one, in fact, had even told me where I was to go, except to the Corozal Sugar Factory: but where, or how far, or in what direction that was, I hadn't a clue.

I went and sat outside, enjoyed the peace and quiet of a glorious Caribbean day - in the circumstances, all rather too peaceful and quiet for my liking - and aimlessly mulled over my non-existent options. Nothing particularly exciting happened for a very long time: and then, at last, one solitary, large, and obviously old American car pulled up into the otherwise empty forecourt. A grizzled, elderly man, dressed in stained baggy cotton shirt and trousers, eventually eased himself out and wandered languidly over towards me. He was, he explained, the local taxi driver. Not much given to celebrating, he'd decided to drive out to the airport on the off-chance that someone recently arrived might be needing his services.

I, of course, not only did, but gladly and immediately accepted them, telling him where I wanted to go. But, I added cautiously, I didn't as yet have any local currency. I severely doubted that he'd be prepared to accept a traveller's cheque, the main form of safe and easily-exchangeable money which we then carried - easily-exchangeable, that was, once the banks and bureau de change were actually open - nor my Amex chargecard, the only other then readily-accepted world-wide method of payment. He might, I hoped, be prepared to do a deal for the few spare US Dollars which I still had on me: because otherwise, embarrassingly unprepared, I was in desperate trouble.

To my surprise, however, he quite happily agreed to take me out to the factory without my having to pay him a thing. Apparently, he said, the 'Sugar Company' would reimburse him later. It turns out that he normally did their taxi work anyway, and someone, who shall remain duly nameless, had simply forgotten to ask him to come and pick me up in the first place.

We set off on our two and a half hour journey northwards. The roads were like nothing I'd ever encountered before: once at least partially tarred, they were now, even in the sparse areas of greatest habitation, so well broken up that they'd mostly reverted to dirt. The required river crossing wasn't any more inspiring, its flat pontoon ferry seemingly constructed of wood and rust, and ramshackle-looking in the extreme. It was also only just big enough to hold a one car and a ten-ton truck: but fortunately, given the day, that was the only traffic awaiting to use it, ourselves included, so we crossed over straight away, courtesy of a man and a hand-cranked wheel, the latter in turn pulling on a long chain lain along the river's bed.

Another hour's bone-jarring drive later, finally back in sight of the sea, we arrived at Corozal's senior staff housing compound: all six bungalows of it. The estate's American general manager, one Mr Neah - or simply 'GJ', as he was always known, for reasons which, along with his real name, I never did discover - was relaxing and celebrating at home along with his wife, and thus more than a little surprised to find me suddenly deposited upon his doorstep. (It again in due course transpired that he should have been pre-warned by that same, careless 'someone'.) Still, Mary soon kindly presented me with an extremely welcome, hastily-assembled meal, and, after the making of a few appropriate enquiries, GJ confirmed that accommodation had indeed already been prepared for me, albeit some three miles away, within the factory compound itself.

It was my first ever overseas assignment for the West Indies Sugar Company, and I was the first Tate & Lyle person ever to work in British Honduras, having been sent out to design improvements to a mill which T&L had, very literally, just bought. They were certainly quick off the mark. I, on the other hand, was far from sure that the manner of my arrival boded such a great start. But I had at least successfully made it to my ultimate destination; if rather sparsely-furnished, and wholly lacking in any further form of food or drink. Without any further ado, I rolled quickly and gratefully into my bed, and awaited whatever Fortune had planned for me next.

Early the following morning, a young local woman knocked on my door. She introduced herself as Rosie, and explained that she was my housekeeper, charged with looking after me during my three-week stay. Things, I thought, were looking up: until she happened to mention that she'd never actually formally cooked for anyone before. To my great relief, however, she added that someone had already arranged for her to be guided by Marjorie Chase, the chief engineer's wife. Their house lay directly opposite mine, and it wasn't long before Don Chase himself, together with his factory manager, Marty Jones, came across to escort me straight over to the boiling house, where I was to work.

From then on, with solicitous concern for my welfare, Marj either oversaw or simply herself prepared every meal which I ate, often alongside the Chases, with whom I quickly became good friends. But the two colleagues who followed me out as residents of the 'guest house', settling in to supervise the work which I'd ended up recommending, apparently weren't so lucky. The Chases had by then returned back to England for a well-earned holiday (as you do when you're working in the Caribbean), leaving Rosie to cook all of the men's meals. Both soon went down with food poisoning: and only then was it discovered that, prior to being engaged by the Corozal Sugar Company, Rosie's main occupation had been that of the local prostitute. What's more, she still went back to her business premises every evening after supper to, as it were, keep in touch with her former clients.

Just think what I'd missed out on! Or perhaps, on second thoughts, not.

I was back in 'BH' by 1966, now tasked with overseeing much of the building, equipment design, and procurement phases for T&L's - strictly, Belize Sugar Industries's - brand new, state-of-the-art, semi-automated raw sugar factory at Orange Walk, about twenty five miles south-west, inland, and up river of Corozal. In practice, I also set out all of the specifications, and performed all of the calculations, for the entire plant's instrumentation and electrical systems, designed almost all of the plant's layout, and supervised the selection and purchasing of all of its equipment. Just a small, everyday task.

The imminent construction works, together with ongoing preparation and planting of their associated cane lands, necessitated laying down many miles of new, white-marl roads. These, together with their dirt-compacted feeder tracks, were absolutely fine, if dusty, for dry weather use: but they turned out to be utterly lethal after heavy - for which, in that country, read perfectly-normal - rain. And so it was that, during my several visits, I learned to drive a Land Rover not only under skid pan conditions, but often leaning over at a crazy angle whilst one spinning side sank inexorably down into the ubiquitous thick, heavy, and very sticky black mud. It was, as it turned out, a skill which would serve me rather well in the years to come.

I also learned a good few ways of successfully cultivating this type of terrain, an undertaking which, perhaps understandably, had hitherto been considered effectively impossible. That probably didn't stand me in too bad a stead for the future, either. Nor did another modern development. Mechanical cane harvesting, generally-speaking, had always been considered far too difficult - and was certainly very rarely encountered, and even less successfully employed - anywhere in the world to date. Yet someone had decided that we, T&L, were going to begin doing it anyway, as much to counter current field labour shortages as anything else. Tower Hill was designated as one of the two chosen trial and development sites, the other being one of our Trinidadian estates, since both were already possessed of extremely competent agricultural engineers. I therefore subsequently spent a lot of time travelling between those two countries, observing their respective problems, and carrying out running modifications to the harvesting machines so that they could not only cut the field cane close to the ground, but also simultaneously cut off its growing top, de-leaf and clean its stalk, then chop this up into more manageable lengths which could finally be conveyed onto an ancillary transporter travelling alongside.

A new form of cane harvesting, of course, then required a new type of cane transport: yet another advancement which fell to my lot to solve. But solve it I did, alongside all of the many other attendant and allied problems, to the extent that, within a year or two, and after visiting several other countries where similar trials were by then taking place, I became known as a - perhaps even for a while 'the' - cane harvesting and transportation expert of those modern, mechanised times.

The new Tower Hill factory was opened by George Price, the country's premier, in early January 1967. I attended my first official inauguration ceremony and formal dinner - although it was, in fact, just a buffet lunch - feeling much like a very proud parent. My offspring's adoptive new owners seemed pretty impressed, too. Bearing "comparison with the most modern in the world", and "breaking new ground" with its processes and facilities - or so said the long explanatory text within Belize Sugar Industries Limited's official booklet - it put the entire Belizean sugar industry "in the forefront" of cane cutting and loading, quite possibly proving to be the "cradle of mechanical harvesting" for the entire Caribbean, Central and South American sugar industries.

Well, the architect of those techniques and equipment - one of them, anyway - would certainly be applying the lessons learnt here for all of those countries' future benefits, that was for sure.

Whilst on my way out for a later general check-up visit, not to mention meeting our newly-installed agricultural engineering manager, I'd been asked to readopt my former factory engineer guise and pay an additional, "as you're already in the general area, loosely speaking", advisory call on a well-known distillery in Puerto Rico. Touring the sizeable plant was in itself educational, both in engineering terms, and as an insight into an industry closely allied to sugar processing, still making good use of the crushed cane after we'd finished with it. Solving their problems wasn't that strenuous or time-consuming, either: so, all in all, it was a pleasant, if brief, diversion.

The plant's management were clearly also very pleased with my recommendations, for, on the point of leaving, one of their directors presented me with three extremely large bottles of their finest, much-sought-after product. These gifts were, of course, most gratefully accepted: but they also presented me with a significant problem, since I already had two sizeable suitcases to carry on my imminent onwards flight. Still, needs must. I managed to fit one bottle into my cabin baggage, and hurriedly stowed the other two within the luggage going into the hold, safely secured deep within, and protectively wrapped with, every possible piece of my normally-neatly-folded clothing.

You can see where this is going, can't you? Picking up my cases as they came off the plane at Stanley Field, I immediately noticed the constant sticky stream dripping out onto the hot concrete. It continued to accompany me all the way though the usual arrival checkpoints - where, most oddly, instead of trying to arrest me for what, surely, must have been some great import-law infringement, given the once-vast amount of illegal liquor which I was obviously smuggling in, the customs inspector simply waved me through without even the slightest thought of asking me to open my bags - all the way out through the foyer, into the waiting car, and onwards to an eventual reception at my host's, the newly-appointed manager's, house.

Impressive start number two!

Fortunately, I already knew Tim Rayvern from our previous work together, so reasoned that I had at least an even change of persuading him that I hadn't suddenly turned into a raging closet dipsomaniac. I'd never, however, met his wife. But she, to her eternal credit, didn't so much as blink an eyelid as this embarrassed stranger apologetically introduced himself. Instead, she simply straight away opened up my sodden suitcases, lifted out their entire rum-saturated contents en masse, and placed them all directly into her large, open-top washing machine. Everything was returned to me early the next day, duly dried and dutifully ironed, and I finally got to change out of my travel-stained clothes.

Helene, as I later learned, was never one to be put out by anything, and I still count her as a very good friend, continuing to exchange letters even after Tim's death.

The contents of the one surviving bottle, by the way, were indeed excellent.

They were a very mixed bunch out in BH, as with pretty much everywhere else that I worked: a group of people randomly forced to live together by reason only of a common type and place of work. During my initial Corozal visit, besides Don and Marj, and then Marty and his exceptionally-attractive girlfriend, Eva, I also soon made friends with Dougie, an agriculturalist, and his wife Jennie. Together, we all often attended social functions in Corozal town; and, occasionally, further afield.

One Friday night, we decided to go for a meal in Chetumal, just a few miles away, but technically over the Mexican border. It was an odd arrangement: the town had been designated as some kind of 'freeport', so no formal entry or exit visas were required to pass between the countries, and official entry into the 'real' Mexico lay somewhere on its other side. Yet you still had to pass through a Mexican border immigration post, its guards conscientiously inspecting every passport before allowing any car or its passengers to proceed: a process which, given that we were all either British or British Honduran, shouldn't have presented any particular problem.

Until, that is, when we were already well over halfway there, Jennie happened to mention that she actually held a Trinidadian passport. That particular country, unfortunately, had recently upset Mexico for some reason or other, and relations were currently strained to the extent that nationals of either were forbidden to enter the other.

"No bother," proclaimed Eva, soon alighting at the border armed with a fistful of passports - not including Jennie's - an engaging smile, an appropriately-alluring hip wiggle, and eyes so mesmerising that the officials didn't even bother to count the total number of people within our two cars.

In no time at all, the governor's band were serenading us from the seafront bandstand as we dined at an open air cafe. Returning somewhat late at night, the border guards just waved us past.

At Tower Hill, it was the chief accountant's wife who always stands out in my memories: an unusually wild character who enjoyed nothing better than 'geeing up' what she considered to be the "old fuddy duddies". The staff houses there were positioned around a wide circular road, in the middle of which sat a few tennis courts and a small swimming pool. Such activities, however, soon began to pall with the lady in question, and so, late one evening, she organised an impromptu car race, involving several laps around the compound's one road. It was, for both participants and spectators, a most exciting event: so much so, sadly, that GJ Neah suffered a heart attack and, before the local doctor could be called, passed away.

The factory's business manager, Henry France, took over as acting general manager until another Henry, Bowden this time, was eventually transferred in from Jamaica as managing director of the whole Belizean operation. (Henry B. then remained in office right until the company was quasi-nationalised; and then, in fact, he remained still, as T&L had been awarded the contract to continue its overall management activities.) I'd already made Henry F.'s acquaintance during one of my earlier visits. He was a widower, I knew, seemingly having lived a lonely existence for many years. He was also generally considered to be a most difficult man to get on with. When he finally invited me over to dinner one night, therefore, most probably out of pure, unavoidable civility, I wasn't at all sure what to expect; I was definitely far from sure that I was looking forward to it.

As if to prove the point, upon arriving at his house, I found him - in the depths of the humid, tropical heat - attired in full British evening dress: black suit, bow tie and all. He was, it soon became apparent, an absolute stickler for proper etiquette, even to the extent of always finishing his meal with a glass of port: not an entirely bad habit, I have to add, although not always the expected or best-matching drink of choice to follow the local, Caribbean-style dishes. Yet it was at just such a stage of the evening that we eventually got around to discussing music. Shortly thereafter, he suddenly asked if I'd like to hear some. In the usual absence of such luxuries during my own temporary overseas stays, I readily agreed: whereupon he showed me to his remarkable, extensive - and, again, not fully what one might have expected out in the middle of the Belizean bush - record collection. It consisted wholly of classic jazz, and it quickly became apparent that he was not only an aficionado, but somewhat of an expert on the genre. Later on in our acquaintance, he revealed that he'd once played drums with some of the most renowned bands which, years before, had visited and performed in top London venues. As it happened, I, too, had already enjoyed my own, if rather lesser moments of glory back in Derby, playing a basic kit, not to mention washboard, in our very local dance and skiffle band.

And so Henry and I finally found something in common. From then on, whenever I visited site, we'd always pass a few evenings together in much the same unusual, but very enjoyable way.

My last Tower Hill trip coincided with the end of Tim's contract. To be clear, I wasn't the cause of it ending - it was as always intended - but I still didn't exactly repay him for his many years of kindness.

It was my own final day there, too, and I'd borrowed his car for a last-minute dash over to the old Corozal mill. All had gone smoothly, and I was well on my way back south by mid-afternoon; smoothly, that was, until, whilst rounding but the gentlest of bends, I unexpectedly found myself diving off sideways and plunging full tilt down a three foot drop, heading straight into the tall cane growing in a road-side field. The supple stalks, to be fair, acted as a wonderful brake, and the car finally lurched to a reasonably-safe stop several feet deep within them, enclosed in a dense cloud of swirling white dust.

Thus covered, and once sufficiently recovered, I clambered back up to the road, managed to flag down a passing cane lorry, and got a lift back to the Tower Hill garage. Gracefully dismissing my abject apologies-cum-excuses, Tim sent out a recovery truck, and soon reported back that his vehicle had actually sustained remarkably little damage, barring having had its fuel tank wrenched off. He'd had it repaired within a few days, I later heard, and all was then, for practical purposes, back in full working order.

I was also reassured to hear that I wasn't to blame. Conditions had been unusually dry for some time, apparently, and the unsupported shoulder of the marl-surfaced road, sitting significantly higher than the surrounding land, had simply cracked apart, given way, and crumbled apart beneath me.

A lucky escape all round, I suppose: if not the most impressive of, this time, endings.

4

EASTERN ESCAPADES

The region popularly known as the 'Far East' was one to which I would always gladly have returned, for a holiday if nothing else. My business travels around the Philippines, Indonesia and Thailand were sufficiently eventful to fill their own separate chapters, and are thus yet to come. The rest, however, still merits a good mention in dispatches.

It was Taiwan which first enchanted me. I'd been asked to present a paper at my very first large professional gathering, the Congress of the International Society of Sugar Cane Technologists. Unfortunately, their 1968 meeting was being held in a country whose existence the UK didn't then actually recognise. Fortunately, the Belgians did, so it was their London embassy who helpfully effected all of the necessary permissions for my entry visa into this geopolitically-troubled land.

And so, with my passport stamped, and my body impregnated with all of the many equally-essential travel vaccinations, I was soon flying off from the recently-renamed Heathrow airport, making the requisite overnight stops in Los Angeles and Tokyo before finally landing at Taipei. To my regret, being purely on a business mission, I wasn't allowed the time to explore either of those two intermediate cities, merely journeying from airport to hotel and back again, and noting little more than how very sophisticated Tokyo seemed to be; a view only reinforced by the fact that they drove on the left.

The preparations made to receive the 500-odd conference delegates were immediately impressive. As soon as I'd cleared Taiwanese immigration, before I'd even reached the airport's main arrivals hall, notices could be seen all over, in several different languages, each directing their respective countries' peoples to appropriate assembly points. I presented myself at the appointed location for UK and US attendees, had my name promptly verified against their list, and was straightway whisked off in a taxi to my designated accommodation. The Mandarin was, as hotels go, very comfortable, very quiet, and, as far as I could judge on such short acquaintance, managed in a most orderly manner. Oddly, though, I soon discovered that it wasn't the same one into which my other two T&L Bromley colleagues, also attending the conference but arriving separately and later, were to be placed: a fact for which I was subsequently to be extremely glad.

The practical and efficient reason for this discrepancy was, I suspect, also the same one behind my own recent lonesome journey, for I had also been booked in to join an additional, pre-conference tour of several of the island's agricultural sites, plantations, and factories. This commenced early the very next morning, with everyone being shepherded into three pre-assigned, language- and country-of-origin-based coaches. Shortly thereafter, we were heading out of the capital on the southbound road, flanked as far as the city's outskirts by a full complement of police motorcycle outriders, complete with screaming sirens and flashing blue lights. Similar accompaniment awaited when we came to the outer edges of the next major town; and so it went on, and so it continued upon the following, and each successive day, wherever and whenever we went. Clearly we were, and certainly felt like, very important visitors indeed; that, or there was something very dangerous going on which no-one had thought to warn us about.

We all preferred the first version.

The island had formerly been called Formosa, or 'beautiful isle', after the description given to it by one of the many boatloads of intrepid

16th Century Portuguese sailors abroad at the time. (These busy explorers apparently had quite a wide vocabulary when it came to such pleasing appearances, it transpires, but weren't exactly terribly ingenious when it came to applying any other descriptors to their various new territories when they first came across them.) For my part, having soon travelled the entire length and breadth of this small, but mostly-mountainous island, with a central, north-south spine of high, tree-covered peaks effectively completely separating one side from the other, I did have to agree that it was indeed a lovely place.

The way that the countryside was organised was also, to us, most impressive. So little of its total land mass was suitable for either arable cultivation or any form of industrial manufacturing that not a scrap could therefore be wasted, every possible available square yard being utilised at its maximum capacity to some form of productive end. Each roadside verge, and even the thinnest patch around or in-between the houses, grew vegetables or rice; the near-vertical hillsides were terraced for tea plantations; soil was being brought down from the mountainsides to infill rocky valleys, so that they, too, could be used for growing crops. I'm not sure where they put the pig farm, but apparently it was very large: and undoubtedly, its by- and waste-products were, like everything else in any given area of endeavour, all most efficiently salvaged and reused. Wherever possible, to maximise efficient crop yields, and thus their own incomes, even the small-scale sugar farmers would be mixing their crops, with rows of, perhaps, lettuce, cabbage or sweet potatoes seen planted amongst the cane. It wasn't going to make harvesting any easier, I thought: but the industrious field workers whom we saw - mostly young women, dressed in conical straw hats and all kinds and mixtures of brightly-hued hat-scarves, blouses, skirts, and trousers - seemed to have it down to a fine art, four of them cutting and stripping off each stalk's roots and leaves, once the cane had been physically uprooted by a male labourer, before bundling them up for removal in ox-drawn carts. They were definitely the most wonderfully-colourful fields which I ever saw.

Not many years previously, a road had finally been constructed to link one half of the huge diving ridge, and thus island, to the other. It followed a narrow, natural cleft, and was still only some fifty feet wide in places, with towering cliffs to one side, and a raging river a good long way below the other. We travelled this road one day - breathtaking in both its scenic beauty and the suspense of its tortuous bends and sheer sideways drops - soon encountering a succession of tunnels and ledges cut into the natural grey, oft-polished marble. All of these, we were told, had been blasted and then dug out by hand: and about halfway along, set upon a stony platform, we passed a memorial to the many men who'd died during the works. Most had actually perished from snakebite, a danger quite alien to us Brits, but very real in these climes.

There was also another nearby plaque, this time commemorating the chief engineer, killed in a rockfall. We did ask our guides who'd pushed the top stone. We didn't, either through discretion, cultural etiquette, or simply a lack of understanding of our humour, ever get a reply.

Alongside an even higher, narrower, twisting side turning, nestled amongst the mist-shrouded mountains, we suddenly found ourselves gazing over a large, placid stretch of water, within which sat a small island, apparently holding special significance for the local folk. Staying the night at a lake-side hotel, we were taken over by boat the next morning, visiting an ancient Buddha shrine and having a local priestess, for a few cents, tell our fortune by picking out a 'lucky stick' from a jar. That visit alone made the long journey worth it, the Sun Moon Lake reflecting an iridescent blue, the whole surrounding setting so typical of the Chinese-type scenes often portrayed in pictures, and upon the popular ceramic plates and pots which so many of our wives and families then possessed.

This certainly wasn't the only antiquity which we saw, either: our tour was clearly not being confined to the business and agricultural sides of the country, as we - and doubtless our employers - had naturally expected. Following the destructive communist 'Red Guard' on the mainland, Taiwan now proudly claimed to be the only remaining centre of 'true' Chinese culture. To prove the point, therefore, not only were we taken to see many old Confucian temples and the like, but we were also given a thorough tour of the Royal Palace Museum. Some of its manuscripts, ivory carvings, paintings, and porcelains, it was claimed, dated back well into the BC era; several colour-printed scrolls and leaflets, produced 'only' in the 12th Century or so, still far predated the first Western printing processes and presses.

Equally impressive were many of the new buildings which we saw, modern edifices sitting quite contentedly alongside their ancient, Chinese-style counterparts. Some, however, often the larger ones, had been very carefully constructed so as to manage the not-insignificant feat of resembling on the outside their grand, traditional ancestors, complete with traditional materials, whilst inside rivalling the latest in US or European designs. The planning authorities must have been terribly proud!

The state of the traffic upon Taipei's roads, however, left a little more to be desired: or, to be fair, was exactly the same as just about everywhere else that I subsequently went, in the Far East or otherwise. Just crossing the road was the usual death-defying feat, even at eleven at night, when the shops were still open and the streets still full of people. Here, though, it wasn't just the weight and general conduct of cars, taxis and trucks, nor the hundreds and thousands of buzzing scooters, and a pretty equal weight of normal bicycles, which presented the main hazards, but the standard Asian tricycle pedicabs and motorised rickshaws as well. It was all, like the cane fields, a wonderfully vibrant, highly chaotic mix.

The one thing which we didn't see a single sign of, however, was the "somewhat erotic behaviour" which I'd been warned that Taiwan was renowned for: both an enticing and highly embarrassing thought for we sheltered types of the time. Nor was there any evidence of the other types of 'hospitality' which we'd expected to be available, in particular the 'home comforter' services which its girls were apparently specifically trained for, in order better to serve its various visitors' specific needs.

Until, that was, we were taken to lunch in the southern city of Kaohsiung. The venue, we were told, was also used for the local "circus and other similar entertainments". Disappointingly, none were actually put on to accompany our meal, but we'd already seen some fantastic balancing and juggling acts elsewhere, not to mention a taste of Chinese opera, along with the more usual song and dance routines to be found at the 'theatre-restaurants' where we ate whilst watching their nightly stage shows. This particular building, though, did seem strangely different. Highly decorated, both externally and internally, its sloping upper walls and ceiling rather resembled Rome's Sistine Chapel, except with one distinct difference: the all-encompassing murals and statues here were not only entirely of a carnal nature, but extremely explicit indeed. It took a good long time to finish our meal, that's for sure, our attentions far more preoccupied by - or, as the case might have been, in trying very hard not to notice - the surrounding decor, and the rather pertinent question of exactly what kind of circus acts, never mind assorted and associated 'entertainments', were normally being performed.

Sadly, we never got the chance to return and find out.

Back in our Taipei hotel, however, I couldn't help but overhear many of the delegates from other 'Eastern' countries increasingly complaining about the unexplained, nay inexplicable, lack of availability of any 'room wives'.

"You know," they informed the rest of us - to which, of course, we innocents abroad nodded sagely - "The ones who look after all the requirements of the room, laundry, making up beds, and all that."

As far as I and my other Western colleagues were concerned, our rooms were always well cleaned, the beds were always changed and properly tucked in, and any laundry was always promptly collected and returned to us, cleaned and pressed, as in any other hotel. Our more local colleagues never did care to enlighten us as to what else in particular we were missing out on here, either; and we, politely, never quite got around to asking.

On the final day before the conference proper, we lucky tour delegates were taken to an "aboriginal settlement" in an extremely remote part of the island. Here, we were told, lived descendants of its original settlers, people who'd preserved and still carried on their old tribal customs. There was a small amphitheatre set up, with a thatched roof and earthen floor: and here, in full traditional dress, the villagers danced us a 'farewell ceremony', its final parts requiring at least the first few rows of the audience - including yours truly - to be decorated with bead necklaces and other part costumes, and then led off by the hand to join in. With reasonably-informed hindsight, maybe this was nothing more than you now get on any modern 'authentic experience' in some allegedly 'isolated, traditional village' just about anywhere that tourists might choose to go. To us, back then, it was highly unusual. What was perhaps the most surprising thing of all, though, was that right at what was clearly the finale, everybody was asked to stand up and join hands. But instead of some traditional folk song, or a spiritual blessing, or some other traditional ritual of some kind, the performers began to sing Auld Lang Syne, firstly in Chinese, then repeated in English.

Maybe not such isolated rural, aboriginal peasants after all, then. Or maybe, to give them the benefit of the doubt, they'd learned it especially to honour us.

The other delegates started to arrive for the main conference towards the end of the week, accommodated across the city in four other hotels. The conference meetings were being held in the Ambassador Hotel, and it was there that we assembled for the inaugural night's dinner, hosted by the vice-president of Taiwan. The mayor of Taipei also welcomed us, proclaiming his city to be the safest in the world, where we could all wander about the streets and back alleys at any hour of the day or night without fear of disturbance or molestation by man or woman. This, as far as we experienced it, turned out to be true: although we later found out from a downtown bar-keeper that the mayor had recently issued a strict edict, covering the entire period of all known planned stays of these important foreign visitors, declaring that there were to be no prostitutes or beggars either on the streets or in the hotels, and that any thefts would be dealt with very, very severely. It had obviously worked; and also, we suspected, explained an awful lot of the lacking, otherwise-expected sights and usual hotel 'facilities'.

Our meal here, like many encountered both before and afterwards, followed the course of a typical formal Chinese occasion, with many courses, and umpteen drinks and toasts. Unfortunately for some, however, despite everything else which had been done in every other way to accommodate and make us feel welcome, the issue of cutlery seemed oddly overlooked. Traditional Chinese implements weren't an issue for me, as I'd used them several times before, and was actually reasonably competent with them. But one of my colleagues, Jan Vandermicke, couldn't manage chopsticks at all. By the end of a week of successive similar official banquets, and just general everyday meals in the country's normal culinary style, the poor, and perhaps over-polite man was absolutely starving.

The conference otherwise went very well, introducing me to many of like-minded interests within the world-wide industry; a valuable resource for the years to come. My presentation also seemed positively received, resulting in several invitations to visit the USA and give similar talks at their meetings. All in all, therefore, except for Jan and the probably-deprived-by-edict 'Easterners', the week was a great success.

Apart for one, far from small matter. Two days before we were due to leave for home, an announcement came over the tannoy system whilst we were at lunch. Anyone staying at one particular hotel, it politely requested (I can't remember the name now, though perhaps it was the Asian?), should please contact the reception desk. Two of my three T&L compatriots, Andrew King and Stephen Gaine, were indeed staying there, but thought that it couldn't really be that important - after all, no-one had said 'immediately' or 'urgently', had they? - and they were certainly going to finish their food before bothering to find out what was going on. I agreed, noting that it was probably just the hotel on fire, and we all thought no more about it: until, that was, we made our way up to the roof terrace for coffee. And there, as soon as we stepped out from the top of the stairs, my joke suddenly turned very sour. The vast column of flame and smoke was more than plain to see, rising to blanket out the entire city's skies. Andy and Stephen beat a hasty retreat. The rest of us could do nothing but to look on and listen to the wail of sirens.

We didn't see the men again until that evening's dinner. Things, sadly, hadn't turned out well for either of them: Andy's had been one of the rooms through which the firemen had forced their entry, blasting tons of water ahead of them, and Steve's was simply burned out. Each had lost all of his possessions, including passport, travellers cheques, and health documents. Yet again, with no formal British representation in-country, assistance had to be sought from the Belgian embassy. They, of course, couldn't actually supply any of the required new documentation, but did prepare very official-looking letters to explain the situation, and arranged for the pair to fly over to Hong Kong, then a full British colony, to receive not only temporary passports, but yet another complete set of inoculations, before finally returning home.

****

Hong Kong was, for me, mostly a mere staging post upon my numerous eastern visits: but I did at least enjoy a few short stays.

I was, at the time, contributing through my church towards the education of a boy who lived there. On one trip, I managed to contact the local agency which was organising it, and they agreed to take me to meet his family. I didn't meet the boy himself - he was, hopefully, at school - but I did take the opportunity to ask the agency's representative if he'd drive me to the nearby border with mainland China. I recall, quite vividly, standing atop a hill, looking over the barbed-wire fence which was there to keep the 'Red' Chinese out: as if it, alone, would ever suffice. I couldn't see how the colony could ever possibly accommodate them anyway, as the local inhabitants were practically living on each others' heads even then. But doubtless that wasn't a primary thought in at least one side's mind.

****

The tiny island, country, and city of Singapore was another major port of call for all journeys in the region, and so was, yet again, often experienced solely through its airport. Still, it did offer the great advantage of being somewhere that one could buy duty-free goods whilst just in the transit lounge, so a routing this way home could always be considered beneficial on those grounds alone. Occasionally, however, I was able to organise, or have enforced upon me in terms of available travel options, the odd one or two days' stays, albeit still affording wholly insufficient time to look around. I couldn't even arrange to meet up with my close cousin Tony, a British naval officer, when his ship happened to be in harbour at the same time.

Of the little that I did see, however, what immediately struck and impressed me was its cleanliness. One couldn't drop so much as a matchstick without incurring the wrath of the local police. Long hair being worn by men wasn't tolerated, either. I once arrived at the airport alongside an Australian passenger who was sporting, in Western terms of the day, a very fashionable style. Without a by-your-leave, he was unceremoniously whipped off for a haircut, before being handed back his stamped passport and granted entry.

But at least I did get to stay at the famous Raffles Hotel. It was actually on my very first visit, a necessary break on that long journey back from the Taiwan conference; long and slow, but also much more stylish and luxurious as a result. Having heard so much of the hotel's history, and how the 'bigwigs' of the pre-WW2 period had always stayed there, or at least taken afternoon tea, it was absolutely fascinating to experience it in person. Much had been renovated after the war, of course, but it still retained the quintessential high, colonial-style ceilings, overhead fans, and a large open terrace lounge for the partaking-in of our said quaint British rituals.

For all of my later visits, regretfully, I had to stay in the fast-appearing modern tower-block hotels: not nearly so interesting, nor, whatever their architects may argue to the contrary, impressive.

The tea didn't taste so good, either.

I was actually asked to pay one, brief visit to a proper sugar refinery, too, situated right within the main city. But oddly, upon phoning from my hotel to make the final arrangements, I found myself not only being requested to arrive between some very specific times, but also being required to gain admission by the not-so-proverbial back door; well, large gate, as it turned out.

Intriguing as this all sounded, it finally transpired, having so snuck myself into the most apologetic manager's office, that the company simply already held an agreement with another consultancy firm: and the owners of that business, it was feared, might well become "extremely annoyed" - the specific outcomes and consequences of which, and for whom, were never overtly clarified - if they ever learnt that a Tate & Lyle representative had even so much as set foot inside 'their' premises.

Suffice it to say that we did gain the advisory contract when it came up for renewal.

****

North of Singapore, across the narrow Johor Straits, Malaysia sits as the tip-and-nail half of a long peninsula which extends down like a crooked finger from the main bulk of Thailand and beyond. Our preferred base when going into the mid or north of the country was the lovely island of Penang, just off the north-western coast. It was a half-hour's ferry ride from the large mainland town of Butterworth, for whose business workers it formed a popular residential retreat, yet you never had to wait more than a few minutes before boarding one of the many multi-decked ships which continually crossed the waters with efficient ease.

Along with a colleague, Bernie Cordell, I was introduced to the island by Mr Gong Li Chok, managing director of Butterworth's sugar factory, by whom we'd been engaged. Not long prior to our visit, Mr Gong's very elderly father, having spent his entire life as a rural peasant who'd never even come near to a city before, had been brought to live out his remaining days at his son's Penang house. Mr Gong Snr. was therefore invited along when we all went out to dinner at yet another Ambassador Hotel. As not uncommon in many Chinese-influenced restaurants at the time, we were suffered, during eating, to watch quite an explicit floorshow, this one complete with transvestite semi-nudes. Both Bernie and I had seen these sorts of things several times before, and, frankly, weren't much enthralled. The old man, however, nearly went through the roof with excitement, repeatedly asking his son if they could go again the next night; and probably several afterwards as well.

Whether his request was ever fulfilled I'll never know: we moved on the next day, and never met either of them again.

Bernie and I were looking into the possibilities of establishing a cane estate and mill in the country's smallest, most north-westerly state of Perlis, bordering onto Thailand. Travelling north for about eighty miles, we made an overnight stay at Alor Setar, from where we were picked up by the state's very own crown prince. He was a thoroughly pleasant young man, to be addressed as 'Tuanku', and, in an old wartime jeep, accompanied only by his driver, he personally escorted us for the entire rest of our tour of possible growing sites.

The countryside was simply magnificent, with huge conical limestone stacks projecting hundreds of feet high up through a flat-topped overlay of dense jungle trees. Annoyingly, this also meant that it was impossible to get any accurate idea of the ground topography. Only much later, when some land was eventually cleared, was the surface actually found to undulate significantly: and yet the trees, for their own unexplained reasons, had all decided to grow their tops to almost exactly the same height.

Having chosen what we thought would make a suitable factory site, subject to soil and other surveys, we returned to Alor Setar, then on again to Penang. What we hadn't known until we'd got back there, though - and whether or not it explained the prince's presence, if not his lack of security guards - was that, because of the strong current threat of invading 'communist terrorists', it was actually forbidden for any 'strangers' to go into that particular part of Perlis at all.

With hindsight, and obviously a good bit of luck on our part, that restriction turned out to be a very good idea. Just one week after our visit, it was overrun by insurgent gangs.

Within a year or so, though, things had calmed down sufficiently for us to be given the go-ahead on a full viability study, followed by a small-scale trial development, for which we appointed an on-site an agriculturalist, Terry Peters. I arrived out not too long afterwards to see how he was getting on, only to be told that he'd recently, and very suddenly, developed a medical problem which prevented him from driving: a bit of an issue, since he was still living in Penang, and making the long car journey to and from Alor Setar and beyond every single day. With fortuitous timing, I was at least able to take over at the wheel for a week or so, by which time it was discovered that his condition was all down to the local coffee, somehow not agreeing with him despite being consumed in considerable amounts. A change of brand quickly remedied the situation, and everything was back on track before I had to leave. Nevertheless, we very quickly found accommodation for Terry and his wife in Alor Setar itself.

I made several more visits to the area afterwards, and also to the Butterworth refinery, whose diverse, multi-racial, multi-cultural management team seemed perfectly to epitomise the entirety of Malaysia at the time: the Chinese generally acted as its overall managers, Indians the technicians, and Malays the general workers. Still, these young men, many of whom had attended colleges and universities all around the globe, from China to India, Germany and England, as well as studying at much more local institutions, all seemed to get on well with each other, at least as far as work was concerned. Sadly, I understand, it didn't always remain so, particularly after the government later became dominated by indigenous Malays, and political differences and jealousies couldn't help but to cause a certain friction, and a good deal of ill-will.

I stayed several more times in Penang, too: but the one place which I never visited - to my knowledge, which none of us, throughout all of our combined stays, ever visited - was the Eastern Hotel in Georgetown. We saw it frequently, a lovely example of the many old colonial-style buildings constructed world-wide throughout areas of former British influence or rule; and, apparently, a very high class hotel, too. Mr Gong Jnr., however, had been quite adamant from the very outset: not only did he refuse to recommend it, but he flat bang would not permit it whilst we were working for his company. The hotel had recently, he'd stated most disapprovingly, started to become what was popularly known as a 'rest and recreation' establishment, and all which that then implied, for the numerous visiting US Navy's personnel. It was certainly not at all, he said, a fitting place for esteemed gentlemen such as ourselves.

His father might well have liked it, though.

****

Sri Lanka - or Ceylon, as I still think of it, probably because I was brought up drinking 'Ceylon tea' - was another former British colony which I visited a number of times. It may not strictly qualify as 'Far Eastern', but it still lies east of India, and these are my memoirs, so I reserve the right to cheat.

My first encounter with the island was along with our group's senior agriculturalist, and my sometime nominal boss, Jerry Sterling. We quickly found ourselves quite fascinated by its traffic. It wasn't, as such, its movement, nor lack of it, but simply the vehicles themselves: nearly every car, van or truck - and there was an awful lot of them in the capital, Columbo - was of an old British make, usually unseen in the UK for many years. Many of the commercial models, in fact, whilst clearly restored into what seemed to be reasonable running condition, had not been fully repainted, so were instantly recognisable as having been formerly owned by the Post Office, British Rail, or one of our several large transportation companies. Or, at least, their various patchworked body panels were: for here, unlike in so many other countries, there were no discarded parts and dismembered shells of old rusting cars and trucks to be seen just left lying around, all available surplus spares being very swiftly, and visibly, used up or repurposed. A year or two afterwards, however, come the mid 1980s, we noticed some more modern vehicles slowly appearing on the roads, and then the standard junked, abandoned hulks did indeed begin to pile up, strewn along about the roadsides of a once-clean countryside.

Thankfully, on this visit, Jerry and I were provided with a comparatively new car, along with a driver and escort. Getting on and about with our usual daily business, we proceeded to travel the island, searching for possible cane-growing sites. We also included visits to their two existing estates, which we unfortunately found to be in pretty poor shape, a considerable part of our time thus unintentionally being spent giving advice on the current plantations, before considering any new ones. Still, our travels did enable us to see a great deal of the inland areas and rural communities, many engaged in logging activities. We found the traffic here fascinating, too, although the resultant slow processions, rescheduled appointments, and general road blockages, diversions, and delays caused by on-site operations weren't always so welcome. It wasn't a mechanical issue, though: it was elephants, forming both their major workforce, and their loading and transportation equipment.

You just don't want to bother arguing with an elephant swinging a large tree in its trunk.

My later visits to the country were in order to draw up plans and specifications for the new factory which had been commissioned following our initial survey, and to advise on the equipment required.

Our team of engineers and process chemists were based at the Colombo offices of the Sri Lanka Sugar Corporation, and accommodated in a gem of another old, former colonial hotel, this one still providing afternoon teas, complete with toast. It was situated roughly halfway between the main city and the extremities of Mount Lavinia - a residential area to the south, blessed with some wonderful beaches, but not a decent slope in sight - and thus benefited from being within easy reach of both, yet away from the worst of the congested traffic.

Pretty much across the road was a large tourist hotel, where we sometimes went for a drink. It was built with a very flat, plain exterior, and a centre which was nothing but a huge open square from ground to roof, into which all of its rooms and their balconies faced. It always reminded us of the archetypal insides of a US prison: although, in this case, with the significantly-attractive addition of waitresses who insisted on dropping down to their knees every time that they served you. We weren't sure whether to take it as somewhat amusing, or just plain, outright embarrassing.

The visiting Americans, of course, loved it.

There's an old missionary hymn which I knew from my youth - it wouldn't go down at all well in our modern times, I'm sure - which referred to spicy breezes softly blowing over Ceylon's isle, where every prospect pleased, and only man was vile. I can't remember the strength of the breezes whilst I was there, I'm not sure that they were so particularly tainted - exhaust fumes often seemed more to be the favoured aroma - and nor, certainly, was everything always so generally pleasing: but I did find its inhabitants to be far from unpleasant.

Interactions with our two native companions and the various industry staff whom we met did, however, throw up a very basic difference between European and Sri Lankan interpersonal communications; more specifically, body language. We were, at first, completely stumped when our escort repeatedly nodded his head up and down in apparent agreement to something or other, but then went on to do, or not do, something with the clear intention of having meant quite the opposite. Even more confusingly, when he shook his head from side to side, it could be either mean 'no' or 'I don't know'; or, perhaps, and often more likely, that he just didn't want to express an opinion at all. The Sri Lankans apparently thought it impolite to show a negative disposition. To the untrained outside - i.e. Western - observer, they never seemed to have a way of indicating positive ones at all.

Amongst the several locals to whom we were introduced was a manufacturing jeweller. Sri Lanka's geology led to it possessing, and producing, an abundance of several precious and semi-precious stones, including agates, amethysts, and rubies. There was also, therefore, a multitude of jewellers, many based in Colombo, many, as to be expected, specifically aimed towards the wealthier foreign visitor. One day, just in a passing conversation, my acquaintance invited me to visit his home and see his wares. Once seated within his perfectly-ordinary domestic lounge, I was utterly astounded as he opened drawer after drawer, and produced necklaces, rings, bracelets, and all other kinds of finished pieces. His prices were a fraction of what I'd seen in the shops, so as well as buying a number of pieces to take home for the family, I asked him to make me a pair of agate cuff links, mounted in white gold. Ready in time for my departure, they were beautifully finished, and I was very fond and proud of them, determined to wear them only on special occasions: until a burglar back home also took a fancy to them, and made it a particularly special occasion by depriving me of any such future pleasure.

The one issue which I didn't come across during my time in Sri Lanka was the long-standing antipathy between the country's northern Tamil peoples and the rest of the majority Sinhalese. In fact, the only example which I can remember at all was the oft-voiced assertion of the secretary of the Sugar Corporation, a man of Tamil origin, to the effect that his background was constantly being held against him whenever promotion fell due. (Whether that was truly a valid claim or not, I have absolutely no idea.) But things obviously did come to a much larger head not all that many years later, perhaps proving that old hymn not so wrong after all.

Politics being politics, of course, it wasn't by far the only issue there at the time. One Saturday, for instance, as a group of us walked as usual back from the centre to our hotel, we were unexpectedly confronted, at gunpoint, by a small group of army personnel. They were manning an obviously-hastily-assembled post: and it most definitely hadn't been there when we'd wandered out. To be fair, we had heard that there might be some kind of trouble brewing - though there were so often such rumours - involving some kind of planned overthrow of the government, currently led by a Mr Jayewardene, by or on behalf of his predecessor, the former (and the world's first ever female) prime minister, Mrs Bandaranaike. According to the soldiers, once they'd conscientiously checked whatever identification we could produce, this time the rumours really were true. They'd been told to expect a large-scale march of Mrs Bandaranaike's trouble-making supporters and other allied 'communists' - one way, or place, and another, there did seem to be an awful lot of them around in those days - and warned us that, in their opinion, almost anything could, and quite probably would, happen. It wouldn't be the first time; nor, increasingly troubled as their country then was, the last.

This, we quickly agreed, sounded quite serious: and so, as instructed, we rushed back to our hotel, gathered up our essential possessions in case of a required hasty exit, and stayed firmly put for several days. Nothing, however, ever seemed to happen. The streets remained perfectly quiet and calm. With no news or evidence to the contrary, we finally decided that it was again safe to venture out, only to find the soldiers all gone, and everything back to normal. Apparently, we then read, for some unexplained reason, the so-called communists had suddenly decided to call it a day, little of whatever was planned had actually ever taken place - at least, not where we were - and Mrs Bandaranike didn't get to take back power. Not that day, at any rate.

5

FILIPINO FROLICS

Our newly-formed international consultancy company had just received its very first commission, carrying out a feasibility study for the Philippines National Bank. They, in turn, were assisting a group of cane farmers to develop and extend their existing holdings on the island of Panay. This 1969 assignment, indeed this entire country, was to present us with a very different impression to the usually-well-ordered West Indian estates with which we'd until then been mostly involved. But at least it was another place where one could - after some kind of fashion - be reasonably easily understood, for, although technically a British colony for just one year, the Filipinos had apparently once claimed that they were the world's third largest English-speaking nation.

Jerry Sterling and I, with the exception of another colleague's brief introductory visit, formed the team's advance party, sent out to prepare for the main work to come. We ended up travelling out separately, with Jerry scheduled to follow me by a few days, so I was met at Manila airport by Tommy Solanas, a man whom I'd first come across at the Taiwanese conference not all that long before, and now representing our local agent, Warner Barnes. The hotel to which he took me was yet another magnificent old colonial-style building, its entrance standing atop an imposing, broad flight of steps, the reception desk centrally placed within a vast, high-ceilinged hall, complete with plentiful rotary fans, and flanked on each side by lifts which transported its patrons to their cool, spacious bedrooms on the three floors above.

Between the hotel and the airport, however, stretched the very non-colonial 'Golden Mile', an area with which, one way and another, I was to become very familiar. The sea ran along the left hand side as one drove into the city, and the buildings lining its opposite length were mostly both surrounded and separated by very large gardens. It wasn't anything particularly special by day, to be sure: yet come nightfall, the assorted decorative lighting and illuminated advertising signs reflected far out across the water, forming a magnificent sight. It was then that the area really came to life, too, nearly all of the buildings functioning as dance halls, casinos, or some such places, many of quality, but some extremely shady.

It was also quite normal for taxi drivers, especially when ferrying passengers from the airport to their hotels, to tout for one or more of these various houses of 'entertainment', more often than not also offering to obtain the use of a girl to keep you further occupied during your stay. These ladies, it was usually claimed, would be a temporarily-unemployed actress, or maybe a TV or film star, singer or similar, and they were always stated to be "very good and very clean". (I hasten to add that I never checked out these or any other related claims in any way whatsoever.) Most of the time, it was just one of those occasions where you smiled in a sagely manner, shook your head politely, and that was that. But some of the drivers could be extremely persistent, not to say insistent, and became quite angry if repeatedly refused. One of my frequently-visiting colleagues, already subjected to far too many such approaches, did, in fact, finally weaken.

"OK," he sighed in defeat. "Yes. If you'll find me a tall, slim-legged actress with long blond hair and sparking light blue eyes, I'll take her."

Fortunately or unfortunately, dependent upon your personal point of view, all Filipino girls, whilst often very attractive, are dark-haired, brown-eyed, and quite short in stature. The outwitted taxi driver just didn't seem to know of any available visiting Scandinavian artistes at all.

I passed my first few days in Manila busily making preparations, meeting contractors and machinery suppliers, and generally learning about their import laws, taxes, fuel costs, and so on. Then it was off once again to the airport, to pick up Jerry.

Having completed his registration procedures, I arranged to meet him back in the hotel's foyer, once he'd had a chance to rest and freshen up. Coming down from my own room a couple of hours later, therefore, I was surprised to find him already well-settled at a table, sipping at what was clearly not his first a cup of tea. I joined him, of course, but he didn't seem at all interested in my company, nor in anything that I had to say. Instead, he just kept staring intently ahead; until, suddenly, he extended a trembling finger.

"Look! Look!" he exclaimed, in his broad Irish brogue. "See that sailor?"

Returning into town, we'd already noticed a large US naval vessel newly-moored in the harbour behind our hotel.

"He came in with a girl, and they've gone straight up in the lift. They didn't even sign in. But no-one stopped them. And they're not the first ones, either. I've been watching for a while now. It keeps on happening, time and time again."

He seemed genuinely upset by what he'd seen, and I had, quite gently and politely, to inform him that not only was this a good old Philippine custom, but that, no doubt, before too long the desk clerk or the bell boy would offer to get him similarly fixed up. I knew, because I'd already refused many such offers during my own short stay.

"I think," Jerry stated, firmly, "That the sooner we get down to Iloilo, the better."

Iloilo was the capital of Panay, itself lying to the south of the main island of Luzon. We were, to Jerry's relief, booked to fly down early the very next day.

We were met at Iloilo by two men: one the chairman of the newly-formed Passi cane farmers association and company, the other introducing himself as Jose Feliciano, owner of the district's largest farm, and, so it soon seemed, also the area's general 'entrepreneur'.

Two functional-looking vehicles awaited us outside the airport's doors: one an ex-wartime US Jeep - which we later christened 'Leaping Lena', since that described perfectly what it did on the local roads - the other a Chevrolet pick-up with an enclosed cab. Each, as to be expected, also came equipped with a driver.

Yet strangely, as soon as we approached, both men immediately leapt out and clambered up into their respective vehicles' open rear beds. It was, therefore, the two senior executives who helped to hoist in our bulky cases, invited Jerry and I to take the front passenger seats, and then themselves drove the next forty-odd miles. Only much later did I realise that the other pair were, literally, riding shotgun: except that what they clutched within their arms were small-bore rifles. And thus, it turned out - if never overtly explained why - we would be similarly guarded for the whole of our visit, wherever we went. Should we ever have need of them, we were also subsequently, quite casually, informed, there were loaded revolvers in the glove boxes, too.

The road out of the city was at first well-asphalted over its whole width. But then, for about ten miles, a smooth tar strip ran only the width of a single car. Beyond there, with the exception of our passage through one small town, it was simply dreadful; even worse than Belize. Not only was there nothing but compacted dirt, but, in some distant past, various people had attempted to improve things by placing rocks and close-pitched stones, rather like cobbles, along several stretches. These 'improvements' had, of course, soon been displaced under the weight of the many large, heavy trucks which passed us in both directions with remarkable frequency, usually hauling cane or some other local agricultural or industrial produce. The consequential, highly-uneven surface currently resembled something like an old Roman route, about two thousand years after its original construction. The many bridges which we crossed weren't much more inspiring, either, made mostly from bamboo and timber, seeming severely lacking in maintenance, and certainly not built to carry modern loads.

Two hours later, considerably internally shaken, we were informed by our hosts that, as denoted by another stretch of asphalt, we were finally approaching Passi. It wasn't a large place, consisting mainly of well-spaced, single storey houses, most with small gardens, a filling station, and several assorted shops. We pulled up in the yard of a good-sized residence, obviously once a private dwelling. It looked quite nicely kept in general appearance: except for the rather incongruous dismantled remains - a complete chassis, wooden cab, wheels, and axles - of a long-disused, and seemingly dumped truck which were strewn around two sides of its parking lot.

Our baggage was unloaded, and we were invited in.

"This is our guesthouse," Mr Feliciano declared proudly. "Please make yourselves comfortable."

I wondered when the near-inevitable invitation to even more comfortably-enjoyable forms of 'entertainment' would follow. Thankfully, it never did.

Beyond the front door, a large room served both as the dining and lounge areas. Leading off from that was a corridor, with two doors on either side. Jerry and I were shown through adjacent ones, finding each leading to a bedroom, both containing two single beds, interconnected by a shared shower and toilet. But there was no time to unpack, wash or rest. Lunch, we were immediately told, would be ready in but a few minutes: and indeed, by the time that we'd returned to the main living space, its huge, highly-polished table was already laid with fine, ornately-decorated china plates, engraved crystal water glasses, and the most beautiful silver cutlery. Its centre was also laden with many covered dishes, whose contents, soon confirmed to be both hot and surprisingly delicious, were served out to us by an elderly, somewhat shrivelled lady who couldn't speak much English.

Finally fully sated, and with the table having been duly cleared, we sat there for a good while longer, listening to the two Passi men talking through their thoughts and ideas, before jointly drawing up a tentative study programme covering all that we wanted to see and do. And then it was straight off out once more, to spend the rest of the afternoon being driven around, trying to gain a general appreciation of the area, and asking a whole load of questions.

Our hosts - either incredibly keen, or just extremely busy - finally returned us, sore, stiff, and very road-weary, back to our accommodation just as it was getting dark. Bidding us a swift good night, they added that Mrs Mangmay - we presumed our wizened housekeeper, since we'd never been properly introduced - would again provide us with dinner. Sure enough, having washed and changed into clean clothes, we returned to find the main room once more beautifully set, with electric lights shimmering from a huge overhead chandelier; flickering more than occasionally, it was true, though thankfully, throughout our stay, we were always supplied with enough power not only to keep the bulbs glowing during the waking hours of darkness, but to maintain the property's all-mod-cons fridges and freezers.

It was another large and excellent meal, after which the old woman quickly cleared, tidied, and washed-up; and then, with but a brief polish of the long, antique-looking sideboard, complete with its own sparkling, cut-crystal bowls, she curtly left the house. A guard still remained on the premises, but, after locking the outer doors, he also fast disappeared off into one of the other bedrooms. Jerry and I remained seated for a while, sunk deep into chairs as well-upholstered as those at the table, puzzling at the unexpected luxuries around us.

"This is some classy guesthouse for a place like this," I observed.

Jerry nodded. "Did you notice that the cutlery's monogrammed?"

Indeed I had: it was all inscribed with a very elaborate 'M'. Only weeks later did we find out that this, in fact, stood for Mangmay: unsurprisingly, perhaps, since our 'housekeeper' and her family, as well as being Jose Feliciano's in-laws, normally lived there. In order to suitably accommodate his business guests, however, Jose had simply and expediently commandeered their home, and temporarily moved them out to share his.

The next morning came as somewhat of a shock. We always aimed to get out into the field before it got too hot, so had arranged with Jose and his henchmen for an early start. Breakfast had obviously been organised to suit. Sometime during the preceding hours, the dining table had been fully re-laid, and was now brimming with dishes of food; dishes which, unfortunately, had also been put out some hours before, complete with their congealed contents of fried eggs, fried bacon, and various other once-cooked, highly-fatty, and now very cold foods.

We contented ourselves with some admittedly-excellent bread and jam, and soon proceeded on our day's work.

At lunchtime, and in the evening - again a splendid feast - no questions were asked, and no complaints were accordingly raised: but the following morning, we found exactly the same. Something had to be done. We tried to explain to Mrs Mangmay, ever so delicately, that, despite being British, we didn't actually eat eggs and bacon for breakfast: it just seemed to be an all-round easier and better way to tackle the problem than trying to organise anything freshly-cooked. And then Jerry, with all of his Irish charm, decided to take matters a stage further, asking a confused-looking Mrs Mangmay if she could kindly show him the cooker. This, he found, also contained a large electric hot plate: and thus an excellent breakfast toast, browned upon a hot griddle, was introduced to the Philippines. It proved, at least with us, to be a huge success.

Another annoying domestic problem soon occurred, though; this time in the middle of the night. Jerry and I both heard it from our adjoining bedrooms: a constant splashing drip from the intervening shower. There was, we quickly discovered, a problem with the valve supplying its head: but, try as we might, we couldn't shut off the water.

"You usually carry a few bits of string with you," Jerry yawned at me. Well, it's the sort of useful thing which the seasoned traveller always carried in those days, wasn't it; that and a decent penknife. "Have you got a long piece?"

I rooted around my various kits, and soon produced about a 6-foot length. Jerry attached one end to the bend of the pipe just above the shower head, and draped the other end down to within an inch or two of the floor. The water flowed down the string, and silently dropped the short distance into the drainage hole. And, lo, no more sleepless nights.

A civil engineer had been booked to join us after our first week. He was flying in straight from Zambia, where he'd been employed on our Nakambala estate. Jerry and I, needing to seek out some further information in any case, decided to go back to Manila to meet him.

I didn't know 'long Jim' Villiers, and, despite Jerry having described him, he turned out to be even larger than I'd expected: well over six feet tall, he was also broad of shoulder, and his legs seemed to take forever to reach the ground. But we got on easily right from the start, and Jim fitted happily into our little team, if not so comfortably into vehicles much better adapted to the small Filipinos. After a short time introducing him around to various local agents, we all made our variously-uncomfortable way back to Passi, where Jim and I subsequently shared a bedroom, and so eventually got to know each other quite well.

Jim quickly adapted himself to our current engagement, often proving, quite literally, to be a tower of strength. Even in the dry season, the Passi area had a lot of 'wet spots', where the soils remained muddy and soaked. It wasn't long, therefore, before we descended a slope on one of our regular field trips, and suddenly found our Jeep sinking up to its axles, just where a stream flowed beside the road. Four-wheel drive it might have been, but nothing that the driver could do was going to move it. So Jim got out, went round the back, and put his shoulder to the panel.

The engine revved, the wheels spun, and the inevitable shower of fine liquid mud sprayed all over him. But nothing much else happened.

"Right," he then said to one of the drivers. "Give me your machete."

He then proceeded to chop down a long sapling, several inches in diameter, pushed it in through the mud underneath the vehicle, and gave it all of his weight. The rear axle lifted fully clear, and, by a combination of leverage, wheel-spinning and pure brute strength, we soon managed to move the Jeep forward onto firmer ground.

Jim, of course, was by this time covered head to foot in a thick, dark coating of slime. Wholly unconcerned, he simply plunged straight into the nearby stream, clothes and all, and washed it off.

After that, as far as the locals were concerned, nothing was too good for 'Mister King Size Japanese', and they'd do anything that he asked.

The name, by the way, without any offence either intended nor at the time taken, stemmed entirely from Jim's preferred choice of clothing. Having come straight in from Africa, he'd brought only the shorts and safari jackets which he normally wore there: an outfit which the Filipinos had otherwise only ever seen on the Japanese troops who'd occupied their country during the Second World War. Such was Jim's influence, however, that a couple of years later, when we returned to implement the factory development phase, he asked if some of his still-favoured safari suits could be made up by the resident Passi tailor. Ever enterprising, the man then not only did an excellent job for Jim, but promptly copied their design and made them up for sale, his garments rapidly being accepted by the locals as their new standard attire.

We were also soon joined by Dr Archie Weatherspoon, an agronomist from T&L's research department. He actually arrived out along with two other colleagues, but they chose not to stay with us very long, casting but the briefest look over the proposed location before arranging for some soil load-bearing tests to be carried out, and then quickly retreating back to the sanctuary of their Iloilo hotel, where they occupied themselves by checking out the port's facilities, and establishing equipment delivery routes. The engineering team weren't, to be honest, terribly impressed. The planned factory was situated within a loop of the river, so we already knew that the soils would be very soft, and we'd factored it into our designs long before the 'experts' had arrived, ensuring that all significant loads were carried on piles.

As it happens, though, those tests did turn out to be a rather good idea: if not for the obvious reasons. We hadn't known it until making contact with the main testing laboratory, but apparently a rival commercial group already had their own ideas about putting up a second-hand factory on the other bank of the river, directly opposite the site selected for the Passi mill. The Japanese, we were told, were also preparing to finance the construction of a new factory at New Frontier, just ten miles away. Still, we weren't overly worried about the newly-discovered competition, as both of these projects, we soon found out, intended to follow the traditional Philippine practice of bargaining with the farmers to supply cane, without undertaking any other involvement. The Passi Farmers' Group, on the other hand, had agreed to follow our suggestion of starting effectively from scratch, including installing a full, badly-needed infrastructure system, allied with the provision of modern transportation and agricultural cultivation equipment, plus the ongoing supply of our advisory services, all being made available for the exclusive use of any grower prepared to contract to supply cane to their new factory. Even the local farmers seemed keen on the new-style arrangements: at least until, like any good farmer, they were requested to release even the smallest areas of their land to allow the new roads and bridges to be put in. Fortunately, this potential stumbling block was swiftly overcome; although not by our ideas, nor intervention. Large hand-outs of cash were apparently involved.

The full survey team returned to the UK after six weeks of work, and proceeded to write up the study's findings. Reports were in due course sent back out to our client, accompanied by the offer of a machinery-financing package arranged by our group's specialist manufacturing arm.

And this would probably be a good time to add a helpful note for the modern reader, for when, here and elsewhere, I use the word 'write' in connection with such activities and purposes, I do, at least for a good part of my working life, literally mean it: as in to pick up a pen - better still pencil and rubber (or eraser, for the Americans) - and a pad of paper, and try to get it right first time. Hand-annotated corrections were generally acceptable, even in final copies, as long as not too plentiful, and done extremely neatly. All accompanying financial and engineering calculations required the additional use of a slide rule, together with books of log and trigonometry tables - the most basic form of calculator only appeared very late in my business day - and many thick tomes of engineering reference data. Technical design drawings necessitated many more specialised pens, pencils, and paper, not to mention many and various 'drawing instruments'.

A secretary, in any commercial setting, was therefore a very valued possession, to be respected, cosseted, and generally treated with all due respect, politeness, and reverence for their invaluable skills; local English-reading and -writing ones, when available, doubly so. Unfortunately, they were not a creature usually found in a form able to be carried around in the bag or pocket, never mind taken out into the field, and thus often had to be summoned up on an immediate and urgent basis as soon as we returned to our main urban office or hotel base, where their skilled eyes and fingers translated and compiled our notes and back-of-the-envelope scribbles - the cassette-taped dictating machine was again a very late luxury for me - into presentable final reports before we forgot what we'd needed to say.

The concept of the plural must also be considered in this context, for, again, there was often no such easily-available thing as a copier. Reports and other documents for dissemination to more than one person therefore either needed prior circulation or, unless they really justified the cost of going to a printers to be typeset and have glossy covers and photographs added, multiple copies manually typing up, and then proof-reading and correcting, over and over again.

Surprisingly - or perhaps not, since the loan terms were very advantageous - the proposals were very quickly approved by the National Bank. Their president, Mr Rosario, joined the various associated company boards, including that of the newly-named Passi Farmers Co-operative Corporation, and our initial study suddenly matured into a full-blown contract for supervision and implementation of the entire set of recommended plans.

The work for this phase was divided into two sectors, a resident project engineer holding prime responsibility for the factory's construction, whilst I acted as overall contract project manager for the 'field' side, as well as handling everything else within the development programme: construction of all planned roads and bridges, the building of six senior staff bungalows and an eight-bedroom guesthouse, and the erection of a large servicing and maintenance workshop for all the various cane, general purpose and personal transportations, along with the various civil engineering and agricultural plant. This not-insignificant remit, its assorted works to be carried out under the watchful eyes of a resident six-man team, with one also acting as field sector manager, and backed-up by my own frequent visits, also encompassed the actual sourcing and supply of all of this equipment and machinery, including tractors, haulage wagons, harvesters, and every other kind of agricultural-type tool and vehicle; and, of course, of the workshop itself, along with its own multiple needs.

Our first two site-based members were Billy Runsey, an agriculturalist, and Alan Draper, our field and transport maintenance engineer. Since accommodation wasn't yet available at Passi, they set up house in the reasonable comfort of an Iloilo hotel. Alan's responsibility was to receive all the new field equipment, tractors, and transport units, and to set them up in working condition. Billy could then start immediate cultivation of the new lands acquired by Passi Farmers for those who'd already signed up to join them.

I went out to introduce Billy and Alan to the co-operative's board, some of whom I was also meeting for the first time. One young man was obviously very affluent, and already owned his own small plane.

He introduced himself as Lucien, the board member responsible for bribing customs officers: which perhaps explains why we were never refused entry for anything, and always cleared goods very quickly indeed through the docks.

My team was soon completed by the addition of Jim Villiers, acting as our chief civil engineer, Ian Vallais, whose duties included the operational direction of all field-based equipment and transport, and Harry Fergusson, senior site engineer, an agronomist by training, and an excellent organiser to boot. We also had an on-board accountant, setting up operational financial systems for both the field and factory sectors. This post's first incumbent, however, a Swiss man, spent only a few weeks supposedly acquainting himself with relevant local practices before apparently drinking himself silly. He was immediately sent back, and replaced by someone from T&L's UK office.

Soon enough, I was back in Manila. Once again, I established residence in the same old hotel, and once more set about the rounds of suppliers' agencies, parts depots, and the like. My allocated budget was well over £1.1 million, to be spent across all of the various equipment, tools, transports, and various construction and infrastructure works. This, for its time, was a very considerable amount of money: under the then-unique terms of the financial agreement, therefore, we had stipulated that all operations would be under a form of 'turnkey' contract, meaning that I, and I alone, controlled all expenditure until everything was handed over on the day of the factory start-up. The contract, as it transpired, was eventually extended, so this financial control continued to cover my remaining site team until the successful end of the first processable crop. Having by then seen all too well the way in which the Filipinos let money trickle through their fingers, mostly in the direction of someone else's greedy pockets, it certainly proved, at least from our point of view, to be a very cost-effective way of handling the available funds.

The factory suppliers had appointed their own resident site manager, one Vinny Groul. He, along with his young family, was initially accommodated in a bungalow just outside Iloilo, and so, just like our field men, had to endure the long, bumpy journey to Passi and back every single day. This really was a pain, and not just bodily: but Vinny soon came up with his own solution, for, amongst his many other accomplishments, he held a pilot's licence. It wasn't very long before he'd secured permission to acquire a four-seater plane, and to have the factory contractor roll out a basic dirt airstrip, later to become the main approach road into the factory yard. From then on, he simply flew in every morning from Iloilo airport, taking him all of ten minutes, and back again at night. There were, of course, a few other added benefits, in that he could also easily fly himself and anyone else around to Manila, or to the neighbouring islands, as often as they required; and he frequently did.

For the sake of the rest of the field team, however, the co-operative had to be pressured to find us our long-promised local accommodations, pending completion of the permanent, on-site facilities. In fairly short order, therefore, Vinny, Jim, and some of the other T&L men found themselves relocated to the 'guesthouse' first used during our initial study. Mrs Mangmay and family had, of course, once again been summarily ousted. For want of a reference address, and as someone's idea of a joke, this residence quickly became known as 'Number 10', the seat of our leader.

Ian, along with his wife and two small children, became the first inhabitants of the second house which we were allocated. This was a large, long, split-level bungalow, with four big bedrooms, one of which I occupied during my visits, and an additional office-style wing to one side. It also had a very nice front garden, centred by a large spreading tree, under whose shade sat a comfortable large bench and table: this was immediately put to very good use for after-work and after-dinner drinks. Acting as our HQ offices, but mainly on the grounds of its external paint scheme, it soon became known as the 'White House'.

Although our influence in Passi therefore always emanated from one of those two well-known seats of power, the real control and internal administration behind and supporting our site and field team still remained firmly and properly where it should be: in female hands. Laura Vallais became the White House's 'mother', supervising the meals, laundry, cooking, and any other necessary domestic matters, whilst, round at No.10, Jim and Harry again had Mrs Mangmay as their major domo for all domestic necessities, until Evelyn Villiers also came out and took over.

The White House wasn't long in use, however, when the team found themselves being continually disturbed, mainly in the evenings, by a succession of random locals. Each would enter the annexe wing through the side driveway door, and proceed to sit, silently and patiently, upon the wooden forms which lined two walls of its outermost room. After a time, the man or woman concerned would usually proceed to bang upon our office's door, and, when someone came to open it, ask where the doctor was, and at what time he'd be back. These particular quarters, it thus quickly became apparent, had of late been the local doctor's surgery. They still would have been, too, had not Dr. Mangmay, another of Jose's brothers-in-law, also been turfed out for our benefit: not only of his home, this time, but also of his business premises, and currently to be found dispensing his essential services from a small wooden hut somewhere on the other side of town.

That alone sealed our determination to start work straight away on the construction of the scheduled bungalows and guesthouse, and it wasn't too long before these contained our full team, the factory erection-engineering supervisors, plus an assortment of visiting engineers from some of the major component suppliers. A guesthouse room was, for the time being, kept reserved for me, and another for the managing director of A. & W. Smith, the factory's designers.

Both No.10 and the White House were then immediately released back to their rightful occupants.

Alan and Billy, meanwhile, seemed quite content to continue

living in the Iloilo hotel. As I also stayed there during my early visits, I got to know the pair quite well, and soon realised that they relished Iloilo's night life, finding that it far outweighed both the inconvenience of their rugged daily commute, and the missing company and sparkling work-informed conversation of their fellow colleagues and respective spouses.

They were, to be sure, an eccentric couple. Billy, for instance, once home from work, took off his field boots and donned a pair of red velvet slippers, which he then determinedly wore for the rest of the night, whether inside or out. And out was definitely their preferred choice. The hotel's restaurant already provided nightly dancing, but their favourite place to eat was at a separate dance hall, which also happened to serve pretty good food. It was obviously a gathering place for the swinging younger Filipinos: yet somehow, 'Uncles' Al and Bill had quickly became popular habitués. Questionably-embarrassing footwear aside, I even sometimes joined them. We usually started out at around eight in the evening, had a drink and a meal, and then, at about ten, I would suggest that, given the next day's early start, we should retire back to our hotel. Clearly mindful of their duties, and respectful of their team leader's wishes, Alan and Billy always readily agreed. Of course, as I very soon realised, after driving back into the hotel's parking lot, they always somehow managed to leave me to go inside first; and alone.

I never found out at what time they eventually returned from their second nightly outing: but they were never late, nor apparently jaded, when setting out for work the next - or probably same - day.

There was nothing untoward or ill-intentioned about their social mixing, though: that was also, and always, perfectly clear. They were both quite happily married, they explained. It was just, they added, that the general presence of so many young and attractive females made them feel extremely "happy with life", and they wanted to make the most of any opportunities presented to them; in their dreams, at least.

Or so they told me.

Our travel problems weren't, by far, confined to the state of the local roads. Politics, it seemed, were just as dangerous as potholes, and some of the villages on our local routes were staunch supporters of the opposing party to that favoured, equally-strongly, by Passi Farmers. It wasn't unknown for the various factions to resort to the use of firearms to deter or otherwise deal with their rivals whenever deemed necessary: hence the shotgun escort during our initial study. We, unlike most of the local inhabitants, did not - and did not wish to - carry weapons: and perhaps for that reason, despite our unavoidably-clear declaration of allegiance, not one of us - with one later exception - was ever directly threatened.

The risks of getting caught up in the crossfire, however, still couldn't be underestimated. One night, for instance, not long after we'd moved our staff to live permanently in Passi, someone had simply walked into the restaurant at their recently-vacated Iloilo hotel and indiscriminately sprayed its dance floor with a machine-gun. It was nothing more than a minor personal vendetta, apparently; a mere flare of temper directed against one of its patrons - the mayor, as it happens - who had reputedly recently 'stolen' the gunman's girlfriend, and on this particular evening was openly seen dancing with her in that hotel. Some of the innocent diners, and the girlfriend, had died. Many more were injured, some badly. The intended target, of course, escaped without a scratch.

This attitude, unfortunately, prevailed throughout the country. Any dispute, whether personal, political, or labour-related, was commonly resolved by use of firepower. One day, the problem came to us, too.

It all started in the site labourers' barracks: three long, wooden huts located to one side of the factory site, just outside a high wire fence, in turn enclosing the checking area through which the men passed at the starts and ends of their shifts. The factory's construction contractor was based over on the 'main island', and had brought in much of its own skilled labour. Other workers, however, came mostly from local sources. Unsurprisingly, there wasn't much love lost between the two sets of men, and, almost inevitably, the regular nightly barracks row eventually bubbled over, guns were drawn, shots were fired, and three men were killed. The scene was thus set for a full-scale battle.

Someone very quickly called out the district's military-style police, and they, also at gun-point, were soon keeping the warring parties apart. But the situation still remained very tense the next day, not at all helped when the local labour gang learned that the contractor's chief engineer was due to fly in to site. This, they very swiftly decided, was absolutely not going to be tolerated until their grievance, whatever it was, justified or otherwise, was resolved to their satisfaction.

And whatever it was, I, at the time, knew nothing at all about it; about anything, in fact. Neither did Vinny Groul. Both of us had spent the previous day and night in Iloilo, deep within several long meetings. Innocently, therefore, and with someone having forgotten, or just not been able, to appraise us of the ongoing situation, we made our planned morning flight back up to Passi in Vin's little plane. Following usual practice, we first buzzed a low pass over the small township, alerting the day's allocated driver to come out to meet us, before circling around to line up with the airstrip, which ran alongside the factory fence.

And then, Vinny suddenly let out a shout: "What the hell's going on down there?"

All over the width and length of the strip, row upon row of rifles barrels, with long lines of men underneath, were pointed straight up at us.

Vin pulled sharply up on the controls, and aborted the landing. Returning to make another wide, cautious circle, this time crossing the runway at right angles, we noticed that our driver had by then arrived, and was standing beside his Jeep in deep, animated conversation with some of the armed men. As we circled once more, we saw the strip being hurriedly cleared, and the next time that we approached, we were able to make an unobstructed touch-down.

By the time that we'd taxied to a halt, the full barricades were back up again.

I never found out how the matter was ultimately resolved, although ultimately - in fact, remarkably rapidly, all considered - it was. What I do know, though, was that whilst we, being involved in no dispute with either the labourers or the contractor, continued to be permitted to fly in and out of site whenever we wished, the 'main island' engineer, under pain of who knows what, in future always had to land his near-identical company Cessna in Iloilo, and then drive himself up.

Oddly enough, even after such serious events, feelings quickly died down, and relationships soon seemed to settle back down to normal. It was somewhat later in our programme of works, in the one exception to which I previously referred, when Ian Vallais ended up in some kind of argument with one of the field supervisors. The latter, of course, having finally lost his temper, did what came perfectly naturally to him, drew his gun, and, with the safety catch off, threatened Ian with the barrel pressed right into his cheek. After the appropriate few moments of tense breath-holding, the supervisor swiftly realised the stupidity of his actions, tucked his weapon back into his waistband, held out his hand, and apologised profusely.

And that, it seems, was that. Excepting, so I'm told, for a very firm friendship springing up between the two men from then on.

Back in the land of the much more mundane - for us, anyway - erection of both the factory and remaining accommodations was soon well underway. All of the tractors and other equipment necessary to commence land preparation on the nucleus estate had been safely imported, and we'd constructed at least ninety miles of good gravel roads and five major steel bridges, along with several smaller concrete ones, all leading directly to the new mill. But, worryingly, the majority of local cane farmers remained most reluctant to sign formal delivery contracts with the co-operative, and we couldn't understand why. Then it was explained: whilst the project's overall concept would undoubtedly prove very beneficial for them, it was also an entirely new approach in the country, and they simply didn't trust that it would ever come to fruition. Therefore, we were told, they wouldn't commit until, as a mark of assurance, they saw the top of the factory chimney being put into place.

And this, indeed, was the case. As soon as the chimney was completed, they all flocked in to sign up as one.

Everyone certainly seemed more than satisfied as we entered the first cropping cycle. Having studied and anticipated various likely local problems, we'd instigated a system to guarantee the identity of each farmer's cane, ensuring not only that it would be processed quickly, but that he would be paid out within a maximum ten days from delivery on the basis of its sugar content. This was something never experienced there before: prior tradition had been for the farmers to hand-load wagons, or in some cases rail cars, and for their drivers then to have to wait at the factory yard - often for several days, sleeping in hammocks slung between their vehicles - until the fast-drying cane could finally be unloaded.

Passi quickly became the envy of the Philippine sugar industry on that score alone.

There'd been initial doubts, of course. Even the company's directors had been utterly horrified when I'd first told them that we wouldn't be supplying a particularly large number of cane lorries.

"But," they'd protested, "We'll be the laughing stock of Iloilo! Everyone has at least 120 big trucks!"

It required considerable time and patience to explain how we were going to use a system by which the freshly-cut cane would be made into bundles in the field, pulled onto the field trailers, lifted by cranes at assembly points into truck and trailer combinations, and then delivered to and unloaded at the factory, all within twenty four hours; and all, therefore, still full of its valuable juices, imminently to be extracted.

It wasn't long before the Passi directors had all set about proudly boasting how they, and they alone, were no longer required to hold such a large and expensive transportation fleet as everyone else.

Not that we always got away with everything. Amongst the tractors which we'd bought were fourteen four-wheel-drive units, designed for heavy-duty tillage. Some time prior to their planned arrival, Mr

Rosario had sent me a message to the effect that he expected to be given one as his perk for getting the overall contract package agreed. I had sent a very polite message in return, respectfully regretting that I couldn't release one because, amongst other things, I held a very limited budget with which to provide all of the machinery and services as outlined in our original study - to which he himself had agreed - and that nothing, most sadly, had been mentioned at the time, and therefore wasn't included in preparing our overall costings and calculations, nor the subsequently-agreed final package, as regards the provision of gratuitous handouts.

The large tractors, once delivered, were immediately put to use. A few weeks later, Alan was checking his servicing records; he could only account for thirteen. No-one knew where the missing one was working, nor where else it could be. Then a small Filipino bird informed us that a nice, shiny, and otherwise identical-looking agricultural vehicle had recently been spotted on, of all places, the main island of Luzon, on a farm apparently belonging to a certain senior manager of the Philippine National Bank.

How strange, we mused, because the model which we'd specially imported was a brand new one, never previously ordered, nor currently stocked, by the manufacturer's Philippines agent.

Anyway, we wrote that one off to experience, wondering if he'd yet worked out how to buy his spares directly from the UK company concerned, because the local branch certainly wouldn't hold any, and only we carried them - now under very careful control - in our Passi stores.

Contrary to our initial fears, the two proposed - later actual - neighbouring factories never presented us with any real problems. The second-hand one, ex Puerto Rico, situated directly across the river, never even got into serious production. The brand new, Japanese-built 'Frontier' factory had plenty of its own worries to cope with.

There wasn't anything wrong with the factory itself: in fact, it was beautifully designed and constructed, as we saw for ourselves when invited to visit. And they, too, had devised a way of overcoming the current long wait, stale cane issues. Unfortunately, it then proved to be their undoing. The planned solution comprised a large number of ten-tonne-capacity bins, to be placed and left in the fields whilst the cutters filled them. Several huge flatbed lorries had been specially outfitted with hydraulic arms, with which then to lift the laden bins onto their decks for carriage to the mill. This may have worked well enough elsewhere, but they'd forgotten to consider the local field conditions - although by no means a problem confined to either the current location or country - where frequent rains left the soils even softer than they already were. Still, they remembered as soon as their heavy transporters entered the first wet fields, and promptly sank straight up to their axles. They then had to retrieve not only the full cane bins, but also the long hauliers themselves. In the end, far from as speedy and efficient as they'd originally envisaged, they simply resorted to dragging the bins to the fields' edges by tractor, before eventually hoisting them upwards and onwards.

Frontier's troubles were doubly, perhaps triply, compounded by the state of the road network itself: in particular, the pre-existing bridges. It was another planning oversight which ended up costing both the Japanese, and many others, dear. The permitted axle loading for their bin-carrying trucks, plus full load, was much greater than anything normally travelling those already highly-questionable and over-burdened public highways.

One by one, one after another, and without any attempt either at, nor contribution towards, maintaining or strengthening them, the lightweight wooden bridges began to fail.

Suffice it to say that they soon weren't the most popular cane processors in town. Not only did the Frontier factory's cane supplies begin to suffer - even sometimes running completely out for a considerable time - but so did almost everyone else's. Except, of course, for Passi's. It wasn't at all uncommon for one of us to drive past a point where one of the co-operative's new radial roads met the 'government road', and come across at least one disconsolate driver, usually from Frontier, slouched down beside his cab, unable to proceed further towards his factory because of a broken bridge. The driver would then ask, ever-so nicely, if it was possible to travel down the 'Passi road': to which we had to answer that, unfortunately, it was not, because the road was privately-owned. It was also, we had to add, controlled at regular intervals by heavily-armed guards. This was perfectly true on both counts. News had quickly spread as to the superior state of the roads and bridges which we'd constructed, and, unsurprisingly, all sorts of both personal and commercial traffic had therefore started to use them, the other factories' cane trucks included. The co-operative's members naturally objected to others using what they were paying for, and, in the typical Philippines way, instead of perhaps bringing in some kind of toll system to the benefit of everyone, had simply placed said patrols at every possible entry point, equipped not only with guns, but with strict instructions to turn away all non-legitimate traffic by any means necessary, no matter what their type or purpose. If even one was let through, it was argued, everyone else would no doubt follow; and, certainly as far as the area's cane farmers were concerned, you either delivered cane to Passi under the co-operative's contract system, or you didn't use the co-operative's roads. If your planned fifteen mile drive to the Japanese factory had to be extended by a diversion of well more than sixty, almost right round the island itself, then, well, you wouldn't have been the first; nor last.

And, by the way, we weren't the ones who caused it.

Someone obviously did repair the existing bridges in due course: but who, and who paid for it, never mind accepted some or all of the blame, I never did care to find out.

The T&L team, fortunately, had quickly established very good relations with all the directors and managers of the Passi project, amongst them another young man who was a close friend of Lucien, the 'customs fixer'. One time, upon hearing that I had to go over to Manila on business, he insisted that Lucien should fly me there in his - or their, as it actually turned out - plane. In reality, of course, this was obviously nothing but an excuse for the pair to go off and enjoy a bit of the capital's high life. Still, it was an easy and enjoyable morning's flight for me: even if, en route, they did insist on flying not only over, but all around the inside of the deep crater rim of, the visibly-smoking Batangas volcano. It was, I must admit, an amazing experience; and, mercifully, not one of those many times when the volcano suddenly decided to do one of its infamously-sudden, unpredictable spouts.

Jesus Garcia was another one of the co-operative's active farmer-directors, owning a very well-managed cane estate, as well as a sizeable farm. Jesus was a great playboy - no, I never imagined that I'd be writing that line, either! - and it was actually his wife, we soon discovered, who handled all of the day to day finances, payroll, administration, and general direction of his entire business interests. Her husband's other pet recreation was cockerel fighting, and in due course, on one of our frequent invitations to visit the farm, he took us over to his yard, delightedly introduced us to every single one of his fighting birds, and explained in great detail how he trained them. We never witnessed an actual public fight, but understood that they attracted large audiences, along with lots of heavy betting. A few Sundays later, Jerry Sterling and I spent a delightfully-lazy morning swimming with Jesus and his family in the local pool - in reality, a large concrete tank which later transpired to be the local village's reservoir of drinking water - and then drove up to his hilltop house to enjoy an excellent roast chicken lunch. We, of course, complimented Mrs Garcia on the quality of the bird.

"Oh," she smiled dismissively, "It was just last night's loser."

Another lady whom we got to know quite well was the wife of the co-operative's company secretary. Once again, I was invited round for a meal - I have never, either before or since, had any problem with kind ladies wishing to feed me up - whereupon she insisted that I visit her basement 'women's workplace'. This intriguing venture wasn't, to my relief, at all what it might have sounded like, for, in the large space underneath her house, Mrs Constantine had established a thriving weaving room. Its three large looms provided well-paid employment for many of the poorer ladies in her village, and, with the aid of imported silk, produced the most exquisite Thai-type cloth. I think that all of our team subsequently purchased and brought home several lengths.

We decided, as time went on, that we needed to start operating a radio network, linking communications between the supervisors' land rovers, the loading stations, and the maintenance workshop. But annoyingly, given the very broken, hilly nature of the surrounding countryside, reception in many areas was patchy to poor. Whilst back in England, I'd held long discussions about the problem with an engineer from Pye, the electronics company based in Cambridge, and he'd eventually offered to come out with some test equipment, aiming to install a VHF transmitter on the highest local point.

His offer was gratefully accepted, and his arrival quickly arranged. The two of us were soon setting out in a Land Rover, along with a pair of drivers, aiming towards the foot of a high mound which rose steeply out of the level plain. The plan was for one of the drivers to take the Land Rover around various nearby localities, checking if and where he could pick up our signals, whilst the other accompanied us, cutting a way through the dense scrub bush with his trusty machete whilst we lugged a portable transmitter, its long aerial, and one very heavy battery up what we reckoned was about 1,000 feet.

It didn't sound like too much of an effort, we'd naively thought, and I'd made sure that we set out early enough to return down before the main heat of the day. And so, slowly, hour after unplanned-for hour, we sweated, toiled, hacked, and hauled our way directly up the face of the slope; and the day just became hotter and hotter, and more and more stickily-humid and airless.

It was already well after noon when we finally achieved the summit. We quickly set up the equipment, and were relieved to find that everything still seemed to work just fine, at least on an outgoing basis. The man below was instructed to drive around in the requisite differing directions, and he soon confirmed from each location that our messages were clear; as, indeed, were his.

We were just sharing congratulatory smiles when, without any warning, the Pye man slumped to the ground in an inert heap. His pulse was OK, so it wasn't a heart attack, I thought: more likely heat exhaustion and dehydration. After all, and perhaps an oversight on my part, he'd only just arrived out from England, and wasn't yet acclimatised to tropical conditions. Luckily for him, I'd decided - if only last minute, mind you - to bring along a large canteen of water, so I managed to get him to drink some, after which he seemed a little better. But we still had to get him back down the hill.

Putting our new radio to its first real, if unexpected, use, I told the roaming driver to return to where we'd started our climb. Then, after hastily packing everything away, the remaining two of us supported - in reality, virtually carried - the stumbling engineer, one on each side, and set off down the path which we'd just cut. Even without our extra burden, descending was much worse than getting up. Three abreast, we slipped, skidded, and tore our arms raw on the scrubby bush as we threaded through. Thankfully, the other driver had the presence of mind to walk up and meet us part-way, so at least he then relieved us of some of the equipment: but it was still well past three in the afternoon when, now utterly exhausted ourselves, we reached the Land Rover and then swiftly got ourselves back to Passi.

My diagnosis, as it turns out, proved to be correct. Our casualty was deemed fit enough to travel back to Iloilo the next day, and thence to fly home with his own tales to tell. A permanent radio station was also later constructed on the top of the hill, although still needing to be attended every few days with fresh batteries. The Filipinos quickly made a much better access track.

My frequent trips to and from Manila, meanwhile, continued apace. Tommy Solanas and I shared many a frantic business excursion around the city, often followed by a well-deserved dinner at some restaurant or other on the Golden Mile.

After one such evening, with Tommy having, as usual - or so I believed - driven home after dropping me off at my hotel, I was woken in the very early hours of the following morning by the ringing of my bedside phone. My initial, sleepy enquiry was met by a brusque, rather menacing voice, apparently calling from one of the Mile's more notorious, if not outright dangerous, casinos-cum-strip joints.

"Mister Solanas was here last night," it continued, addressing me by name for good measure. "He lost a lot, and now we have him in one of our rooms. You must bring the money he owes us." A very considerable sum was then named. "We won't let him go until he's paid up."

I didn't for a moment believe that Tommy was the sort to have left his family and gone out on a late-night gambling spree; nor any other type of nocturnal activity which one might associate with such venues. Someone, I decided swiftly, was distinctly trying it on.

I asked to speak to Tommy, but was flatly denied.

"If you don't bring the money, it will be all the worse for your friend."

Ah. Extortion with added threats, too.

The problem was, I only had Tommy's office telephone number, not one for his house; and there wasn't, to my knowledge, such a thing as directory enquiries, even if I had known his address. I simply couldn't check whether or not he actually was - as I strongly suspected - at home. Yet I'd been given a very strict, pre-office hours time limit by which to take the payment, after which something potentially rather unpleasant would apparently befall him. I was, just in case, rather reluctant to take the chance.

It wouldn't be any good trying to phone the police station, either. They were always late turning out, even for traffic control, and at this hour of the morning, experience also told me, they'd be far busier changing shifts.

So, a game of 'Call My Bluff' it was. Having dressed, I went downstairs and saw, in front of the hotel, a taxi whose driver I'd used several times before, and whom I felt, within reason, that I could trust. What he thought of my subsequent strange behaviour

I'll never know, but he never even raised an eyebrow as I asked him to take me to the closed casino.

The large building was set back from the road, within in its own grounds. I told the driver to carry on right up to the entrance steps, and to park there very visibly and ostentatiously. He was to remain in place for fifteen minutes, I stressed, and mustn't leave his car. If I hadn't come back out by then, he should immediately go to the police station, no matter what shift or break they might or might not be on, and make one heck of a fuss.

Then I gathered my thoughts together, drew myself resolutely up to my full five foot seven-ish inches of height, and entered the lavish casino's marbled hall. The lights were burning brightly, but all was deathly quiet.

"Hello," I called out in a loud voice. "Who's there? Anyone around?"

I had to shout out quite a few more times before an extremely sleazy-looking gent finally appeared. I deliberately didn't go towards him, instead staying just within the entrance doorway, letting him approach me. I then told him that, in answer to his telephone call, I'd come to collect Tommy; but added, in what I hoped were suitably-firm tones, that I'd intentionally brought absolutely no money with me at all. I also warned him that if both Tommy and I weren't back outside within fifteen minutes, my driver was calling out the full might of the city's law.

I watched his facial expression change from the ready expectation of receiving his blackmail money to one of incredulity, and thence to stupidity. For a moment or two he hesitated, not sure what action to take. Then he mumbled something in the local Tagalog language, before saying, in rough English: "Oh, we changed our minds. He's already gone home."

"Right," I responded. "He'd better have, or I'll be back. With company."

Turning on my heels, I strode briskly and haughtily out to my taxi, and requested to be returned to my hotel.

Needless to say, as soon as his office opened, I rang to check that Tommy was indeed there: which he was, having just enjoyed a sound and refreshing night's sleep, safe within his own bed. It was only then that I suddenly broke out in a cold sweat, realising what could have happened had I appraised the situation wrongly. Pure bravado, or pure stupidity? I certainly hadn't stopped to think much about my own safety. Far too many airport spy thriller novels...

I never did find out how the casino had got my name and room number. I suspect that someone on the hotel staff knew of my business connections, and had passed the information to the casino's owners in exchange for a cut of the profits.

Moreover, though, I never was, nor ever felt, threatened again.

That particular incident aside, my Manila visits were always, in varying interpretations of the word, entertaining. Each of the various restaurants along the Golden Mile had excellent little bands - 'combos', as they were then known - and often also evening 'floor shows', ranging from comedians to rudimentary not-quite-fully strip shows, where the lights always went out just at the last moment. But what I really enjoyed were my lunches with Warner Barnes's manager, Martin Wise. He was a member of the British Club of Manila, a place so typically a replica of a London gentleman's club - well, from my purely imaginary capacity, anyway - that, once within its doors, it was difficult to imagine being anywhere else. An oak-lined hall led into an equally-oak-lined library, its shelves filled with the great British classics, its tables covered in scattered copies of Punch, Field and Sport, and the Tatler. There was, of course, a magnificent dining room: but we never actually went there to eat. Instead, we made our way up the wide stairway, complete with its huge, hanging chandeliers, to the grand billiards room. There, resplendent under traditional overhead lighting, was a full-sized table, with tiered sets of polished and cushioned spectator benches covering three sides of the room. We not only had a game or two whenever we visited, but then partook of an excellent buffet lunch, personally attended by our own, white-coated steward.

Oh, how the other half lived!

Life for the full-time ex-pats could still be pretty difficult, though. There were signs everywhere of corruption, extortion, and physical fear. Most of the Brits whom I met lived in their own enclave, a vast area entirely surrounded by a stout brick wall, not only tall, but then topped with several strands of thickly-barbed wire. There were only two gateways, each having sturdy wooden, equally-high gates, always kept firmly shut except when their guardians were sufficiently satisfied that the person outside had the right to enter.

General standards of responsibility and accountability were also, to say the least, somewhat lax. About a year before my first study visit, the city had suffered an earthquake. I'd already noticed that some of the old, apparently-irreplaceable, colonial-style tiles were missing from my hotel room's bathroom wall: shaken clean off, so I was told. Then one of my business acquaintances, a prominent civil engineer, took me for a drive around the city, pointing out some of the more major damage sustained. One particular, sizeable property had been constructed with an open garage for its ground floor, the six residential floors above supported upon rows of stout columns. Their bases had all simultaneously collapsed, and the entire building had thus lowered itself straight down by full single storey. Many of its former inhabitants' flattened cars were still just visible beneath the front walls. Yet amazingly, no-one had suffered even the slightest injury, and the block was once more partially in use.

The Ruby House (or Tower), however, had not fared so well. All but its top two levels were now nothing but piles of dust and rubble. About five hundred people had been buried during its collapse, several hundreds had subsequently perished, and my companion had sat on a subsequent panel of enquiry into the disaster, so was fully cognisant with all the facts. The supporting columns here, he explained, had been - should have been - made out of re-enforced concrete, the casting process efficiently designed so that their positions could be encased, the steel re-enforcing bars inserted, and then the concrete poured in, two floors at a time. The metal rods and wires were then to be left protruding from the top, Greek-style, for the architect to see: which was all very well and good, except that he seems to have been so keen and swift in his examinations that, each following night, the builder would come back, pull up the bars before the concrete was fully hardened, and reset them in place prior to pouring the columns' next stage. When the quake struck, therefore, all of the lower layers, entirely non-reinforced, had simply disintegrated into tiny pieces, literally exposing the evidence for all to see; or, in this case, not.

Said builder, it was generally understood, had immediately 'gone missing'. No-one knew to which country he'd fled with his ill-gotten gains, at the expense of so many lives.

I soon witnessed these kinds of problems in practice for myself, too. Late one night, wailing sirens could be heard coming from all directions, converging on a spot not far from my hotel. The city's university - to be precise, its separate women's section, including their dormitories - was on fire, thick flames already tearing up through the roof. Then I noticed a succession of fire tenders queuing in the gardens fronting my window, filling their tanks with water; water which emerged from the stand pipe in such a slow trickle that the firemen had plenty of time not only to get out of their vehicles and stretch, but to walk around, have a chat, and even smoke a cigarette or two. It seemed that, for no explicable reason, the fire hydrants in the university area had long since not carried water, and no-one had ever really bothered about the fact - more likely, never even bothered to check - until they came to be needed.

Come daybreak, and the buildings were completely gutted. All of the university's female students immediately had to be sent straight back to their homes, and there they would simply have to stay, uneducated, for as long as it might take - weeks, months, probably years - to acquire or construct their new facilities.

These kinds of attitudes and outcomes aside, there was, to my mind, no doubt that the Filipinos were very ingenious people in many, many ways; and not just in 'manipulating' money or goods in kind. The Passi locals, in fact, seemed to be able to make just about anything; and out of just about anything, too.

Except decent roads and bridges, apparently.

Outside our original 'guesthouse', as previously noted, lay strewn the once-associated components of an old truck, discarded like all the other piles of junk seen scattered here, there, and just about everywhere else. To our amazement, however, just before commencement of the cane-cutting season, the heap in Mrs Mangmay's yard was suddenly seized upon by one of her sons. In a burst of frenzied activity, he soon had the chassis members jacked up on old oil drums, the springs, axles, transmission, and wheels were attached underneath, an engine and gearbox was mounted at one end, and a large steering wheel slid into place. Finally, the whole lot was encased within a home-made wooden bonnet, cab, and rear canopy section, and fuel was poured into the tank. Lo and behold, the engine fired up at the first turn, and the reconstituted truck rolled straight out onto the road to commence duty as a cane transport unit.

Apparently, we were told - although the explanation did cause a little quizzical head-scratching from our perspective - all such trucks were routinely dismantled at the end of each year's harvest as a way of preventing theft, as well as preserving their major components in good working order.

Many of the locals' houses were by this time built of what, to us, were conventional materials: but some, especially for the general labourers, were still made of bamboo, one of the more readily- and easily-available local materials. Iloilo's giant bamboo indeed lived up to the name, with stems averaging four inches in diameter, and forming useable lengths of at least twenty feet. These provided not only the houses' main supporting frames, but could also be split down into narrow slats which were then woven to form both their sides and roof, the latter then thatched with palm leaves. The floor, laid out about six feet above the ground, was similarly constructed, allowing a flow of cool air up into the house from the shaded region below, whilst sliding, split-cane 'windows' allowed light in during the daytime, but could be closed for privacy at night; and not a single screw or nail was used in the entire building, everything being fastened together with bamboo pegs.

This type of construction had another advantage, too, becoming particularly obvious if and when the occupier wanted to relocate. Bumping our way along a road one day, we were surprised to find it entirely blocked by a suddenly- and certainly rather inappropriately-sited bamboo dwelling. We were sure that it hadn't been there the day before: and it wouldn't be there the next day either, for, upon closer examination, it proved to be mounted atop a trailer which was being pulled slowly along by a water buffalo, the standard draft animal for those parts. They were, quite literally, moving house.

Bamboo was also used for so many other things, amongst which were drinking vessels and huge water carriers. The latter were like long-handled barrows, with a large-diameter stalk for the water container, and one or two attached wheels cut from a cross section. Pipes of hollowed bamboo were commonly used for irrigation, including in the cane fields, and, halved, they made excellent rain gutters. Once, on Luzon, I even saw a full musical pipe organ which was made entirely out of bamboo. Sadly, I never got to hear it played: it must have sounded wonderful.

This most useful of plants was also, as already mentioned, one of the main construction materials for all the local roads' stream and river bridges: but they, for fairly obvious reasons, also as already stated, didn't always fare or function so well.

The island's transportation system wasn't only confined to roads and bridges. Its single track, north to south railway had been built by the British many years before: and had never, it appeared, rather like most of the roads and bridges, been touched since. Apart from the wagons which used the line to deliver cane to some of the mills, there was also an occasional passenger train. Having seen the unevenness of the actual rails, far from parallel in places, never mind their original, now non-existent gravel bed, I had grave doubts about ever using one of these: but Harry Fergusson was nevertheless eager to give it a go, and somehow persuaded me to accompany him on a trip to an older sugar factory, some twenty miles to the north.

The train departed Passi with a sudden, uninspiring lurch. Our carriage immediately began to proceed with a disconcerting kind of crabbing motion, as if one corner always moved ahead first, only to be caught up and passed by the other. The passengers' seats might, in some far-distant time, once have been described as upholstered, and were set directly upon a plain, wood-planked floor; a floor which, I saw to my horror, was now opening up and then quickly closing, plank by adjacent plank - just a crack, about an inch or so wide - at first one side of the coach and then the other. The whole floor was alive beneath my feet, moving in a spasmodic dance to the deformity of the rails along which the train's wheels ran. I stared down, mesmerised, through one gap and then another, confidently expecting the entire thing to fall to bits at any moment, leaving us with our feet dangling above the sleepers and rails which, all too plainly visible, were speeding past beneath.

And yet we arrived at our destination wholly intact, and spot on the designated time; and subsequently, if with only a little trepidation, also safely completed our return trip strictly to the stipulated schedule.

On reflection, I thought, it was overall probably a much less stressful, and certainly more comfortable journey than had we made it by road.

Amongst my many project management duties was the making of arrangements for our site team's families - had they chosen not to join their husbands and fathers on a permanent basis - to travel out to Passi for the school holidays. This usually included Billy's wife and four children, Alan's wife, and the two oldest Villierses. The latter, a boy and girl aged ten and twelve, attended two different UK boarding schools, and, to complicate matters, rarely had identical holidays. It wasn't therefore unusual to end up accommodating one or the other for a day or two at my own house, so that they could travel out and back together. Still, it wasn't exactly the most onerous or unpleasant of my tasks, and there were never any major issues, for those two or any of the others. All that I had to do was to book the flights, send a messenger to the Philippine embassy in London to collect their pre-arranged entry visas, and have our company driver transport the children to Heathrow, where the airline's 'aunts' (staff specifically employed to escort unaccompanied travelling minors) would accept them into their care until safely delivered to their parents at Manila airport. Everything was nicely covered: or so I'd always thought.

And then one day, only two weeks before the end of the autumn term, with the pair due to fly out to site for Christmas, both of the Villiers children's schools independently kindly thought it time to inform me that the youngsters' passports had just expired. To complicate matters, although their parents carried full UK passports, the children had been born whilst Jim and Evelyn were living and working in South Africa, and for some reason currently held only passports issued there.

Frantically, I phoned the Home Office in London.

"No," they told me, in brusquely-official terms. "Terribly sorry, but the children cannot now acquire British ones."

So I turned to the South African embassy. They were even less helpful, merely supplying me with the various forms required to effect the renewals: long, complicated, and requiring an awful lot of additional documentation.

Fortunately, the children already had their birth certificates and other papers with them in England, so a driver was sent out to each school to gather them up. Unfortunately, the forms also required the signatures of both parents: and they, even via the medium of the fastest possible international air mail service, would certainly not be so easily obtained. I finally managed to get Jim on the telephone out at site, far from as easy or quick a process as it might nowadays sound, and arranged for both him and Evelyn to fly straight away to Manila in order to meet a special form-carrying messenger whom we were at that very moment in the process of dispatching with all due haste.

Two hurried handshakes in the airport's arrivals lounge, two quick signatures scrawled upon two pieces of paper, and one sleep-deprived courier immediately turned straight around for the next London-bound flight.

Two now-complete wads of forms sent straight over to the South African embassy, and two new passports issued on an urgent basis.

All appropriate papers and documents finally collected together, fully filled in, collated in the required order, and personally delivered to the Philippine embassy just in time for them to issue the necessary visas.

One huge sigh of relief from yours truly.

Job done.

Or not, because the Philippine government, like many others, had just adopted a 'ban apartheid' policy.

Oh joy!

After a good deal more time-consuming telephoning, not to mention a great deal of grovelling and pleading, their embassy staff finally relented; a little. They would issue the visas, they declared, if - but only if - the children were prepared to sign a letter saying that they didn't agree with the apartheid system. I severely doubted that the youngsters as yet had the capacity or knowledge even to understand much about it, never mind the ability to express such a valid, legally-binding opinion: but in the interests of practical expediency, tongue firmly bitten - and just grateful that they hadn't asked the parents to do it, too - I was able to persuade their similarly-minded schools to arrange what was required. For all that I know, maybe the Villiers and their fellow classmates all benefited from an impromptu lesson in current geopolitics.

Whatever the truth of it, and with only one full day remaining before the children were booked to travel, two very carefully-worded undertakings were handed over, a crisp, clean page within each of two shiny, brand-new passports was embossed with a small smudge of colourful ink, and off to the airport the young pair happily went.

And I, at least for that year, finally relaxed.

Passi's development phase was, by this time, complete. The factory was due to start full operations in the following January, although the site team was engaged to stay on for one more year, not only to see the first crop through - including supervising its loading and transport, and checking the maintenance of all equipment under operating conditions - but ensuring a successful first re-ploughing and planting of the nucleus estate. This, however, was to be my last visit; and I still had to undertake one more strange encounter with my local friends.

The Filipinos were, notionally at least, intensely religious. In that part of the world, this meant belonging to the Roman Catholic faith. Passi possessed one large, stone-built, and, regretfully, very dilapidated church, regularly used for Sunday Mass, but not for very much else. There was also an immense cemetery, its many graves often accompanied by the most ornate headstones and statuary, all enclosed by a high and wide stone wall, along whose inner face were set a number of wooden doors, each sealing the burial place of the township's earliest residents, entombed within the wall itself. Yet the departed here were most definitely not forgotten. All Souls' Night, November lst, was a big occasion in the village calendar, the time when one customarily spent the night with those who had gone before. This particular year, not knowing quite what to expect, we'd been invited to join them.

It was dusk when everyone gathered at the church, the locals, great and small, all carrying an assortment of large parcels, bags, boxes, and folding chairs. Candles, torches, and lanterns were lit, and a priest then led the procession over to the normally-peaceful cemetery. The old grey walls and gravestones, some new and white, some old and stained, were soon quite beautifully illuminated by a mass of flickering, softly-glowing lights. There began a short and suitably-respectful religious ceremony, praying for the souls of those interred within; and then a loud general hubbub broke out from all around. The chairs were opened, locked into place, and placed around the graves. The boxes were unpacked, and linen cloths were spread over the grave-tops. Out came the food, along with many bottles of beer, whisky, and wine. Once suitably and heartily fed and 'watered', mellowed but increasingly noisy, the villagers proceeded to produce cards, dice, and other games: and in no time at all, a rip-roaring, night-long party was in full swing.

It was a wake to end all wakes. Even the Irish would have been impressed. But they all still turned up on time for work the next morning, sore heads and all.

On my very last night there, in a balmy mid-December, Jim and Evelyn Villiers invited me to dinner, along with Harry and Michelle Fergusson. The site team's bungalows had been built atop a little mound, just outside the main village. Their front patios afforded a wonderful view over the cane fields to distant hills, whilst palms and other tropical trees grew in the gardens amidst a mass of colourful flowers. So it was there, after eating, that we sat and relaxed, each of us silent and thoughtful, with music from a record player quietly in the background, and no light other than that from the stars above.

And then, through the darkness, we caught sight of a dim, wavering light, far away at the foot of the slope. We weren't expecting anyone, and unannounced visitors - in fact, any form of traveller at this time of night - were quite unusual. The light was, in any case, far too small for a truck or car, and was approaching far too slowly for anyone riding a bicycle. It was, we realised, as it drew closer and began to climb the hill, a single candle lantern, swinging from the top of a pole. Moving nearer still, the weak pool of illumination revealed it to be carried by a small group of local children, one oddly clothed in a long, blue and white dress, another wrapped up in a thick brown cloak.

They soon neared us, stopping just a short way off.

And then they started to sing.

I don't know when I've ever felt so moved, hearing Once in Royal David's City sung by such a group of innocent youngsters, the Virgin

Mary and Joseph included, under a clear, tropical night sky. Was this what was like so long ago in Bethlehem, I couldn't help but wonder?

They then sang another carol, their thin voices carrying on the warm evening air, wished us all a happy Christmas, and turned to wind their way back down to the village. I don't know if we gave them any money, but I expect that we did; my eyes were too filled with tears to see them go.

It has to rank as the most wonderful, beautiful, and heart-warming send-off which I have ever had.

I needed that memory, as it turned out - not to mention a bit of divine support - on my own way home. Harry, ever the adventurous and intrepid traveller, had decided that he'd personally come to see me off at Manila airport.

"Do you know," he'd remarked suggestively a few days previously, "I've never been on the overnight ferry from Iloilo. What about going back by sea?"

I agreed that it could be interesting, and was certainly different: so we booked onto the afternoon departure from Iloilo's port.

Arriving early at the ship, we were firstly shown to our cabin, then returned to stand on deck and watch the other passengers board. Few of them had paid for berths, so most just found themselves a space wherever they could around the decks, upon the side gangways, once retracted, and amongst the open cargo. We set sail in the fading light, and then, having watched the last of the land slip slowly out of sight, Harry and I made our way to the near-empty dining saloon, ate a none-too-appetising meal, washed it down with a couple of local beers, and turned in for an early night.

The ship was already starting to progress with a distinct rolling motion, and, as we lay restlessly in our bunks, its movements continued to increase. It was soon obvious that we were ploughing through a full-force tropical storm. I could only wonder how the 'open-top' passengers were managing, no doubt by then feeling pretty seasick, not to mention getting very, very wet. But it had blown itself out by the time that we rose the next morning, and we ate breakfast whilst entering the shipping channel for Manila harbour, with everyone safe and sound.

I've often thought of that trip, though, for, about a year later, the very same ship was lost, sunk along with all on board, whilst making an identical journey in probably pretty similar weather.

The rest of my overnight trip back to the UK passed without further incident: until, that was, I entered Heathrow's customs hall, whereupon, quite unusually, I was stopped and asked to open my cases. Being only a few days before Christmas, I was carrying a good many beautifully-wrapped presents which I'd offered to bring home and deliver to friends and relatives on behalf of my team. I didn't have precise knowledge of their contents, but I'd been given a general idea, along with a note of their value (in those days often of as much concern as the nature of the actual item), so the stop was merely an annoyance rather than any particular concern.

But an annoyance it definitely proved to be, the customs officer insisting on opening every single one of the parcels, and examining each and every constituent part; and then, duly satisfied, simply stuffing everything randomly back into my cases - wrapping papers, silks, ribbons and all - and slamming their hard lids shut.

Perhaps he was a little put out that he'd found nothing suspicious. I was a little more than put out when I had to spend the whole of my first day back at home trying to re-wrap the various gifts before hurrying down to the Post Office to dispatch them onwards to their designated recipients, along with suitable, hastily-scribbled apologies for their condition.

It wasn't by any means only the foreigners who gave me problems.

6

INDONESIAN IDIOSYNCRASIES

(or should that be idiocies?)

It was the late 1960s, not long after the overthrow of the Sukarno regime. Indonesia's sugar industry had once led the world: yet of late, we knew - and as I sadly witnessed in so many other countries, too - it had fallen into a severe decline, and was now in urgent need of resuscitation. I was therefore sent out to see what could be done.

This time, though, I wasn't travelling as an adviser or consultant, but as a pure salesman, peddling our services to their government. And when I say 'our', I don't mean just those of Tate & Lyle, although I still went out under their name. The proposed study of sugar processing operations, present and possible future, covering the entirety of the widespread and very numerous chain of islands which comprised the Republic, was of such a large scale that it would require a sizeable joint team drawn not only from several T&L companies, but from Bookers and the British Economic Commission as well.

All local arrangements had already been made for me by the Ministry of Agriculture and Industry, including the provision of interpreters - in the main, a group of young students eager to catch up with the wider world's use of English - and I didn't have any specific or structured plan to work to, over and above paying my respects to the minister and then following through whatever he thought necessary to prepare for commissioning a more thorough investigation.

Once collected from the airport, I was quickly installed in a small hotel a short distance from the centre of Jakata, itself set at the far north western tip of the long, slim island of Java, the former capital of the old Dutch East Indies 'empire'. Newly-constructed, the hotel building comprised only two storeys, its ground level containing the usual coffee shop, restaurant and bars, and the upper one all of its rooms. Three years later, I stayed there again. By then, it had gained a further four floors.

My first visit, unsurprisingly, was to the national sugar authority's offices. Then it was on to inspect one of their local mills. And then the minister decided that I ought to examine the area around Surabaya, at the far north east of the island, where the greater part of their plantations and factories lay. There was, however, still some civil unrest in those parts, which remained under military control. I must therefore, the minister decreed, travel at all times in the company of an army officer; and no lesser one, I found, as we were swiftly introduced, than a full general. Annoyingly, I also discovered, he knew not a single word of English. But, I was reassured, another interpreter would be provided as soon as we reached at our destination, if not during the actual journey itself.

So off we both went to the airport, and, simply by following the general's gestures, successfully boarded a local flight. Landing at Surabaya, we were promptly driven, now complete with full military entourage, but still not one English-speaker, to another hotel. The entirety of our route, right from the airport's exit into the very centre of the city, was lined by armed soldiers; no unauthorised movement of civilians was apparently allowed at all. I hoped that they hadn't gone to all that trouble just for my - and the general's - arrival.

This hotel, which flourished under the name of the Bristol (probably once having had British connections, I suspected), was even smaller, looked extremely basic, and its little proprietor, who seemed to do absolutely everything himself, also spoke nothing but Indonesian. The general and his subordinates promptly disappeared off to their own barracks, and the promised interpreter had yet to put in an appearance: but eventually, with the time-honoured exchange of a few odd 'foreign' words, and many more gesticulations, I managed to ascertain the times during which meals were served, and then made my way to my somewhat strange room.

In one corner, there sat a square, concrete structure, about three feet high and the same in width. There were no visible taps, but it was already filled to the brim with water. Alongside lay a wide, solid slab, upon which stood a sort of saucepan, albeit fitted with an unusually-long wooden handle. That wasn't the only odd thing, though, for, upon looking at the bed, in addition to the pillows in the usual place, I saw a long roll of what I thought must be bundled-up sheets.

"Ah ha," I reasoned, "It is the local custom for the occupant to make up his own bed."

Anyway, as it was already late afternoon, and I was both travel sore and dusty, I decided to spruce myself up. There was only one obvious solution, so I climbed in over the wall of the concrete pool, spilling out the water in my best Archimedian impression, and enjoyed a cool but refreshing bath.

When I recounted this tale some years later to an old colonial civil servant, he smiled rather knowingly, before explaining that my bathtub was merely a large receptacle, and that the correct procedure, more akin to a shower, was to scoop out and pour water over one's body with the 'baler' provided, whilst standing on the slab. Oh, well: there weren't any signs or instructions, I wasn't to know any different, and I can only hope that no lasting harm was done. Mind you, I do hope that they changed the water before the next guest came. Perhaps, with hindsight, before I had, too; I don't know how long it had already been standing.

Dinner was difficult. I couldn't order anything, as the menu was written, by hand, in Indonesian script. Still, something reasonably edible was given to me - I hope that they took pity on my innocent Westerner's plight, rather than standing watching and laughing at whatever horror they'd really presented me with - and, after eating, not being able to converse with a soul in the communal restaurant, I took myself straight off to bed.

A bottom sheet was, I found, already fitted. But there was no top covering. I thought that I'd better unravel something from the large roll: only nothing that I tried revealed a way of peeling anything off. Upon closer examination, in fact, it actually seemed to be nothing but some kind of long, padded bolster. No matter: it was warm enough to sleep uncovered, and I presumed that was exactly what the locals did.

My latter-day informant subsequently also told me that my roll-cum-bolster was actually what was known as a 'Dutch wife', to be placed, in times of need, alongside oneself and cuddled - in lieu of a teddy bear, obviously - as you slept.

The next morning, I faced the same, inevitable problems with breakfast. Still, I was again presented with something which didn't taste too bad, and, fortunately, it wasn't then too long before my new young interpreter finally arrived, along with our now singular, senior military escort. I politely decided not to mention the unconventional - or, presumably, with hindsight, perfectly normal - sleeping and washing facilities, but instead, duly cleaned, rested, fed, and able to communicate properly, gratefully got down to the business which had actually brought me there.

Over the course of the next few days, I saw a very good number of extremely antiquated and thoroughly clapped-out factories, not to mention some very unusual systems of field preparation and cane planting which, frankly, seemed to me like they'd come out of the Ark. The reality, of course, was that they'd simply existed since times long past, and to date had well suited both the conditions and requirements of the Indonesians' lands and culture.

I also saw an awful lot of something else which, often near as antiquated, still clearly suited their needs: their - to quote a postcard which I kept from Surabaya - "unique traffic". Wherever one went in Indonesia, although particularly prevalent in Surabaya and Jakarta, the streets were always chaotically crowded with the ubiquitous 'becaks': a reversed, pedal-powered tricycle, whose two-wheeled front consisted of either a passenger cab or a space for carrying foodstuffs and goods. With only the most occasional 'modern' - so again, mostly highly dilapidated - car or small truck competing for space, these rickshaw-cum-pedicabs formed the majority of the inhabitants' main modes of transport by far.

Our travels finally brought us to the country's Plant Research and Nursery Centre. Known to sugar technologists everywhere as 'Proefstation Oost Java', or 'POJ', this experimental facility had previously been regarded as the world's foremost site for the development of new cane varieties, along with methods of growth and treatment. I'd been particularly looking forward to going there, I must admit, although at the same time deliberately preparing myself with the expectation that it wouldn't now be as it once was, back in its full glory days. Even so, never could I have anticipated the utter neglect and dereliction which faced me. All of its glasshouses were in ruins, every single pane broken, and everything which wasn't hung with huge spiders' webs was covered in thick-looking mould. There was virtually no instrumentation left at all, and what little else remained was not only ancient, but almost certainly useless.

I left the general behind, and returned alone to Jakarta, utterly depressed.

The following day, after listening diligently to my gloomy observations, the minister proceeded to ask me a good many questions, including what I'd recommend as a remedy. And so there I was, once again providing the full ad-hoc advisory service, whether engaged to do so or not.

In the end, all indications were that he would award us the proposed contract for a full-scale survey.

"But there is one thing you should know," he added. "There will have to be an included fee of fourteen percent."

At least this time he was up-front and open about it.

"But this is not all for me, though. Oh, no! We have a scale." He then proceeded to detail exactly how the added 'mark-up' was to be split, right down to the relative pittance allocated to one lowly ministry official.

Once back in the UK, I reported my findings and opinions - and the minister's additional requirements - back to my employer, and hence their consortium partners. The survey contract, in whatever form found satisfactory to all sides, was in due course signed, and the work took place over the following two years: but, because of other commitments, I wasn't to play any further part.

One of that study's recommendations, however, was that the same joint group of companies, together with the World Bank, should carry out yet another set of investigations, this time to ascertain whether new sugar estates could be developed in a number of other identified locations throughout the wider archipelago. Three years later, therefore, I returned as part of an eight-man team, headed by Jerry Sterling.

Or, rather, it should have been. We were all set to fly out on from Gatwick on a Tuesday morning. Our very tight schedule had again already been fully arranged by and with the ministry, all hotels and transport pre-booked, our outbound and return flights - unusually utilising the relatively new, and doubtless remarkably cheaper, charter flight industry - fixed as to times and dates. And then, at home on the preceding Sunday, I received a phone call from Jerry, regretting to say that he was still stuck over in Ireland, having fallen and badly broken his ankle whilst at his sister's farm. If all was well, though, so the hospital thought, he should be fit to fly out and join us in about ten days' time, once the swelling had subsided and a new plaster cast was sufficiently hard. In the meantime, I'd have to take over the leadership role.

The rest of the T&L team comprised Doug Roundes, an agriculturalist and emergency Jerry substitute, Clay Scott, our most aptly-named soil scientist, and Sean Mannering, a factory engineer. To them was added Wesley Doe, one of Bookers's agricultural engineers, an Australian economist called Colin Massey, and two men from Wallace Evans, a civil engineering company, named Stewart Redman and Robert DeCoste. None of the T&L men had ever met Stewart, Robby or Colin before, so it was a bit of a gamble as to how we'd all fit together when we met at the airport for the very first time.

The answer, happily, was well enough; and, one short overnight stay in Singapore later, we all arrived in Jakarta.

The first two days were spent either at the ministry, or off gathering whatever information, maps, weather data, and other sundry knowledge we wanted before heading off to our first assignment on the large island of Sumatra, immediately to the north west. Then we flew out over the short straits separating the two main Indonesian islands, looked down on the extinct remnants of the infamous Krakatoa volcano, and landed at a small tarmac strip in southern Telukbetung. Our appointed guides were waiting to meet us outside its little wooden terminal, quickly loaded all the baggage and other necessary gear into their three Jeeps, and soon had us on the four hour journey north to our base town of Palembang.

Returning there a few evenings later, I was handed a brief telex. Jerry was due to arrive in two days' time. Having already arranged our week's field programme, none of us could be spared to go and collect him, so we instead sent one of our drivers back to 'Teluk' to meet the flight. It wouldn't exactly be difficult to recognise this particular passenger, we told him. After all, especially in those days, at a tiny airport in a relatively unusual destination for Western travellers, how many other white, middle-aged men would be hobbling off the plane with the aid of crutches, sporting a lower leg encased in white plaster?

Once the driver had eventually disentangled this most amazing of coincidences, and with due apologies all round, the correct man was delivered up to us in good time for supper.

Jerry, of course, wasn't yet up to field work, so we put him to use as our base director for the rest of the Sumatran studies, paying any required in-person calls on local persons and institutions, and generally seeking out whatever further information which we deemed necessary. The rest of us carried on exploring our own sizeable project areas, until it was time for all hands to return to

Jakarta, and prepare for the next destination on our whistlestop tour: Hutan Loyang.

Hutan, we were told, meant 'forest' in Indonesian, this extensive part of western Java having once been some kind of forestry zone. Additional efforts had, apparently, occasionally been directed towards small-scale crop cultivation - albeit long since abandoned - but, until then, there'd been no attempts made at growing cane.

Our hotel was in the northern coastal town of Cirebon, set upon a railway line which ran east from Jakarta, then turned south down to the historical royal city of Yogyakarta (or Jogjakata). For a change, therefore, instead of flying down, we and our interpreter boarded a smart-looking, and very comfortable train, and set off for a ride of several hours, during which a uniformed attendant served us with frequent free cups of tea, often accompanied by something to eat.

We were joined at Cirebon by the local field agriculturist, Daniel, who'd already arranged the supply of three Jeeps and drivers. With no rest for the wicked, these stood readied to set out at first light the next morning. Leaving Jerry to write up our last trip's outstanding notes, the main field party travelled south for about an hour on a good tarmac road, finally turning off onto a dirt track which soon reached the second of the five preliminary cane-growing trials which we were to see. Having examined its results to date, we took a general look around the area, but could see little evidence of any other activity except for dozens of women and children emerging from the scattered groups of slender trees and their surrounding general tall scrub, each carrying heaps of branches for firewood. So much for the forest.

We quickly determined that we'd have to look a little further off the beaten tracks. Daniel, with his usual willing efficiency, offered to arrange three suitable motorcycles: and so, extremely early the next day, six of us were driven back to the trial site, from where we set out along the local paths on three very small bikes. Doug rode on the back of Daniel's, I took pillion behind Wes, and one of the civil engineers drove the third, whilst Clay, behind him, protectively cradled his long, bulky, but ever-ready soil auger - in effect, a hugely-oversized sharp, spiral drill bit - with which he intended to bore and extract sample cores. It was a rather a comical sight, to say the least, but it certainly made access to the terrain much easier, and we were quickly able to get a good feel for what was there; a very good and close-up feel, it must be noted, given how many times we all fell off.

We still couldn't find anything which seemed suitable for growing cane, though, and our allocated time there was fast running out. After further examination of what little, wholly-inadequate mapping there was to hand, we agreed to extend our search a good deal further afield. Like as not, we'd only have the one opportunity for such a foray deep into these uncharted territories, so we decided that the whole team - minus Jerry, his leg not up to such a long day's pounding - should go out together in the three Jeeps, along with our local drivers.

And so, equipped with a good supply of food and water, we once more assembled at dawn, fully expecting, despite the miles to be covered, to make it back by early evening; definitely in time for dinner. We picked up Daniel, drove straight on past the trial plot, on past the land which we'd explored on the motorcycles, and eventually turned down a narrow dirt track, along which the twelve of us, tightly-packed into our three vehicles, then made good, if rather uncomfortable progress. Clay dutifully called a halt every now and then in order to take a soil sample, but for the most part, hour upon monotonous hour, we saw nothing but scrubby, dried-up bush to either side.

And then, in the very middle of the proverbial nowhere, we came across a small group of huts. In-between them and the track's edge, if just for three or so yards, ran a shoulder-high fence made of densely-woven bamboo, oddly but suitably obscuring whatever it did. Soon enough, though, as our noisy procession approached this tiny, remote community, undoubtedly accustomed only to the slow passing of a few animal-drawn carts, its purpose became all too clear. Shouting and waving enthusiastically to us as we drove by, the several curious young women, eager not to miss this rare spectacle, seemed to find no difficulty at all in first scrambling up and then sitting on top of the surprisingly-sturdy screen behind which they'd just been partaking in their usual morning's bath. We, of course, being terribly polite gentlemen, waved merrily back as we regretfully continued on our way, somewhat bemused, but altogether delighted at the full splendour of the sights just revealed to our very appreciative gaze.

Towards midday, we came across a small, dried-up river bed, the first which we'd seen that day. It was reasonably narrow, yet surprisingly deep, and fortunately was already spanned by a short, flat, and decidedly uninspiring-looking set of wooden planks. Yet it did at least prove that other wheeled traffic had, at some unspecified past time, already progressed further down the track: so much so, in fact, that it was now quite rutted and eroded, here lying well below the level of the bordering land.

With its passengers making the crossing on foot, the first of our Jeeps proceeded to attempt, and indeed safely succeeded, to pass over the bridge; a bridge which, it was plain to hear, was more than somewhat loose in the decking department. The second Jeep crossed rather more cautiously; and then the third started off, its driver more confident now, and consequently a little faster.

The several dislodged planks fell into the stream bed with a sound remarkably like that of a wooden dulcimer's scale.

We weren't unduly disturbed, though, as the Jeeps could still, according to their drivers, quite easily make it down the banks, across, and up out up the other side come the time for our return.

So on we went.

And then, not all that many minutes later, it started to rain. Not just any old rain, but a torrential tropical downpour, soaking us all to the skin, and immediately turning the sunken track into a full-blown cascading waterfall, constrained by the higher land on either side. The rear Jeep swung straight back, only to return and report our previous route no longer serviceable, its once-dry gully already full of deep, rushing water, far too dangerous to drive through.

If only there was still a serviceable bridge...

We waited around for a short while, but finally, with no sign of either the rain or wash-down abating, realised our stark choices: resign ourselves to staying here the night, and maybe longer, or hope to find another way back.

We pressed on.

The rain poured down until late in the afternoon, when it stopped just as abruptly as it had begun, and the sky cleared into a beautiful late afternoon. The track had also continued straight on for just as long, without any sign of possible alternatives. At last, however, and now at least able to see a little more clearly, we came to an odd level area, where a long, slightly-raised mound stretched across our path from right to left. This, one of the local drivers said, if memory served him right, might be an old disused railway track which once, somewhere, connected into the main line: and indeed, as a closer inspection soon revealed, so it was, albeit that its rails were now missing, and its remaining sleepers had mostly sunk into the old earthen bed.

Consulting our rudimentary maps, we agreed that there was at least an indication that there ought to be some kind of railway, or remnant thereof, somewhere in the general vicinity of where we might well now be. (The unmarked track along which we'd been travelling all day, we later found out, had actually been put through many years earlier to service the line when it was in use.) Furthermore, the drawings confirmed, it should lead back in the rough general direction of Cirebon. Well, it was the best information, and option, which we had. And at least, logic said, if we followed it then we should eventually meet with a road, mapped or otherwise, and thus be able to find our way back to some kind of civilisation, somewhere.

All went well for a few miles, as we drove along the relatively-easy grassland beneath and parallel to the extinct line. But then the embankment began to rise higher above the surrounding ground; or, more correctly, the land fell away into dense, impassable scrub. There was nothing to do but to go back and find a place where the vehicles could climb up to, and then travel along, the old raised railway itself: a process which proved not to be too difficult or uncomfortable an undertaking - albeit not anywhere as fast as we'd have liked, given the already-failing light - since the wooden sleepers had sunk so deeply into the ballast that we were virtually running, if in a rather bouncy, jolting kind of way, on the flat.

Then we came to another narrow crossing over yet another narrow gully. This bridge, complete with its not very many, and certainly not very sturdy-looking, cross-members, was supported at each end upon some kind of masonry; exceptionally-crumbly-looking masonry, we quickly deduced. No-one much fancied our chances. But, after disembarking and inspecting the water levels, the lead driver reported that this particular obstacle wasn't actually as steep or deep as the last one, and, as the water levels were also beginning to fall, it was probably safe to drive through. Without too much trouble, therefore, and just a little slipping and sliding, the three Jeeps in turn made their way down and back up its banks, whilst the rest of us walked carefully, one by one, across the bridge and joined them on the other side.

Things, we thought, were looking up. Good, if slow, progress was once more resumed, as we bumped our way along the haphazard old track in the gloomy dusk. But then, suddenly, the leading Jeep's brake lights flashed on, and we all juddered to a stop. Ahead, illuminated in the headlights, we saw our next challenge: another wooden bridge, but much longer this time, spanning a deeply-incised, good-sized river, some thirty feet or more in width under normal conditions, and even more swollen now. The ground also sloped down towards the gorge, so that the deck of the bridge - what little there was of it - ran considerably higher than, and so carried on for a good distance beyond, the mere span of the river's banks: perhaps a hundred yards in total. And, of course, being a railway bridge, it required no solid decking, the old rails - here still in existence, in places, if decidedly rusty - being carried only upon their remaining sleepers, themselves directly supported along two main timbers which ran between the bridge's twin trestle piers; sleepers which, by the way, also displayed some very distinct spaces between them, below which there was nothing but a large, gaping space and some very fast-flowing water.

Someone had previously laid a very crude form of walkway beside the rails, but only to one side, and consisting of nothing more than a thin line of parallel bamboo stems, logs, and small tree trunks, helpfully augmented with a rickety bamboo handrail. All once fine for a train, of course; and all still fine, in theory, for an intrepid local pedestrian seeking a short-cut.

Not at all fine for three Jeeps.

It was fully dark now, if with a clear, starlit sky. Once again, we discussed our options. Wait it out and try again tomorrow; carry on walking along the track; turn around and somehow retrace our steps, hoping that we could manage to re-cross that original steep river bed, once sufficiently re-dried. We wouldn't starve for a while, although we'd already eaten most of our provisions for lunch, and we certainly wouldn't die of thirst in the present conditions: though whether any of the stream and river water was safe to drink, never mind just to get down to collect, was a question which we really didn't want to put to the test if we could possibly help it.

Then one of the drivers had a bright idea. If a sufficient number of logs were laid out in a suitable crossways pattern either side of the rails, rather like the adjacent footway, he could drive at least one Jeep's length forward at a time, and then, by a process of leapfrogging, the supports lying behind the vehicle could be picked up and placed back down in front, until he reached the other side. We all agreed that this, in principle, was a sound idea: except that, before we could start to repeat the whole process for the second Jeep's run, the entire collection of logs, trunks, and whatever else we could find to use would physically have to be moved back again from one far end of the bridge to the other. Twice, in fact.

In the end, we decided that we might as well carry out the whole operation as one, allowing all three vehicles to travel together. The drivers, as usual, had long cane knives (or machetes) with them, and so, after an hour's hard labour, taking it in turns as we worked by the light of the headlamps, we'd chopped down all of the saplings which once grew alongside the track. They were, at best, just a few inches to a hand's width in diameter: but, once laid across the sleepers, they were enough to form two makeshift decks, each about a foot wide, and the length of just over four Jeeps.

Our initial trial was successful, and the lead Jeep travelled one whole length forward, the following drivers keen to proceed - slowly this time, please! - close on its tail. All that the passengers now had to do was to walk back and forth around and alongside the moving vehicles, picking up and putting down the slender, if not exactly lightweight 'logs' as we went, whilst taking great care not to overbalance or lose our footing and fall off the bridge at the same time. It was worrying enough if you were working on the walkway against the bamboo handrail; to the other side, there was pretty much nothing underfoot at all, and in place of the absent handrail only a vertical drop, and plenty of pitch-black void through which to fall.

What could possibly go wrong?

Pretty much immediately, we struck trouble again. Without any form of lashing to bind and secure the slim, rounded trunks, the first Jeep's wheels had inevitably sent them rolling them apart and aside. As soon as the second vehicle started to move off, one of its rear wheels slipped halfway down into the now-exposed, and just perfectly-sized gap between the existing sleepers: and there it wedged firm. There wasn't even anywhere to brace a jack underneath, so it was simply a case of human hands doing all of the lifting: a task made even worse not only by the lack of space to put one's feet, but the very real risk and consequence of ourselves slipping through those insecure supports, or, for half of the party, falling backwards over the unprotected edge.

Still, we got it done, the offending lengths were once more put back in place, and on we went. Until, of course, it happened again. Standing in the darkness upon either a very thin and extremely-dubiously-constructed walkway, or else upon the significantly-spaced-out sleepers, labouring under nothing more than the occasional beam of a head or tail light - in turn ruining any night sight which you'd just accumulated, and providing a multitude of shadows and blind spots in which to bump into each other, or worse

- we once more lifted the wheel back up and relaid the decking; and then repeated, and repeated, and repeated.

Once, nearly at mid-river, I failed to see one of the gaps, and one foot, and then entire leg, went straight down between two sleepers. I clutched at the bamboo handrail before me, only for it to collapse in my hands as I pitched forwards onto my remaining knee, saving myself only by arching my body forcefully, and throwing all of my weight backwards onto, and clawing and clinging for dear life at, the stuck Jeep in question. Several hours passed - or so it felt - before someone could make it round to hoist me back up, more than a bit scraped and bruised, but thankfully with nothing broken or otherwise too badly separated.

I'd never told anyone about my not-so-long-forgotten childhood fears and nightmares, the mesmerised terror which they'd evoked when crossing certain bridges, usually over water, and usually of the slatted, see-through type. The ones which always ended with me somehow slipping down between the planks, then swept away by the raging torrents below...

It took us several more hours to cross that interminable bridge, inch by cramped, cautiously-creeping inch. On the way, we worked out how to use some of the logs as levers, and learned how to reduce the hazardous congestion by working in turn in separate gangs. By the time that the first vehicle at long last reached the far side, shortly followed by the other two, we had it down to a fine, if utterly exhausting routine. But finally, we were all able to reboard, and began to bump our way off again along the significantly-more-solidly-secure track towards... Well, right at that moment, we really weren't too sure that we cared.

Several miles further on, we spied a small, hard-surfaced track running beneath the rail-bed. Soon enough, we were hacking our way with delirious abandon through the thick bushes which blocked our descent. We didn't know where it led, and could only guess which way to turn once we reached it: but it had the look of a proper, intended-as-such strip of roadway, and that was good enough for us.

Half an hour later, we were back at what we recognised as the main route into Cirebon.

It was well past two in the morning when we reached our hotel; over seven hours later than our estimated worst-case time. No-one was there to greet us with a shared relief. No-one was pacing the hallway in anguished consternation, worrying where we were. No-one had even left a light on for us, never mind anything to eat or drink. We had to bang on the doors and windows in order to rouse a distinctly-grumpy porter to let us in.

After a long, restorative night's sleep in his dry, cushioned bed, Jerry casually enquired at breakfast if we'd discovered anything useful. Having - in-between yawning, stretching and massaging our various cuts, bruises, aches, and pains - explained all of the difficulties which we'd encountered, and all of the trials which we'd undergone - cue more rubbing and wincing - we then, with the humour which follows such things, proceeded to tell him all about the bathing beauties who'd greeted us from behind their screen.

At this point, Jerry sat bolt upright, and became extremely agitated.

"You ought to be more careful!" he burst out, suddenly looking very concerned indeed. "That could be terribly dangerous!"

Dangerous? Us?

"You know, in some of the Javan tribes, it's mandatory for a man, if he sees a naked woman, to have to marry her."

Where he'd got that bit of information from we couldn't guess: but we assured him that we hadn't left any calling cards, and also, since we hadn't actually even slowed down, never mind stopped, they would, so to speak, have to catch us first.

"Of course," we joked, "That's why we took the long way home. Just to be sure. And we were very careful to have our tracks washed away, too."

Some people, of course, are gluttons for punishment. Clay took his profession very seriously, so was terribly disappointed at not having collected samples from the areas through which we'd mostly-unintentionally passed. He therefore decided to go back and study them again.

Setting off a day later, he and Doug asked Daniel to drive them as far as the broken plank bridge, from where they, with Clay's cumbersome auger slung over his shoulder, started to walk. They didn't return at all that night. Just what, we desperately wondered, had happened to them now? There were thoughts of bottomless ravines, raging torrents, and ravishing, husband-seeking women; and everything in-between. Any and either way, even if they weren't lying somewhere, injured or worse, or wandering around, hungry and hopelessly lost, they couldn't speak a word of Indonesian, so couldn't easily summon assistance or ask for appropriate guidance back even if they did find someone to help. We resolved to go looking for them if they hadn't returned by the following morning.

Come early evening, however, and the pair calmly sauntered back into the hotel, informing us that, contrary to all our worries, they'd passed a very pleasant night sleeping in a spare hut in a small, friendly village which they'd come across - all females fully clothed, and helpful communication surprisingly possible - and had easily walked back to the trial area the next day. Clay was so happy at having gathered the required number of soil samples that he even ceased his usual grumbles. For a while.

This was, with some considerable cheer as far as I was concerned, the end to the Javan part of our trip. We packed up and returned to our Jakarta hotel, reported back again at the ministry, then flew onwards once more.

Our next investigations were based around Ujung Pandang, on the south-western coast of the multi-armed island of Sulawesi (or Celebes), much further east. (As a general point of interest, this city was also known as Makassar, from which derived both the tar-like oil which passing sailors sold to eighteenth century men for the dressing of their hair, and the associated English term 'antimacassar', cloth coverings for high chair-backs which otherwise became somewhat greasily-stained.) We were accommodated in a ten-bedroomed hotel, the entire original eight team members, plus Jerry and our single Indonesian counterpart, for once each enjoying the luxury of a room to themselves.

There was, of course, a downside: the proprietor, in his halting English, apologised for having no provision to feed us. However, we were pleased to hear, he had arranged for us to take all our meals at a restaurant immediately across the road. What we didn't at the time appreciate was that it served entirely Chinese food: which was fine, and very nice, for a few evenings, but became a little more than trying on our unaccustomed palates and digestive systems over the entire two-week stay, morning as well as night. We did at least get some relief by buying bread and various fillings to make up lunchtime sandwiches, and, fortunately, soft drinks, beer, and bottled water were in plentiful supply.

There were three more experimental sites here, each tended by a young Indonesian agronomist. We were not only to examine the cane at each, and to survey the surroundings for likely new cultivation lands, but to scout out a new factory location. Back in yet another trio of Jeeps, we went first to study the site in the Takalar region, just to the south of our 'UP' base. This daily journey necessitated the crossing of a magnificent, not to say wonderfully-solidly-constructed bridge, its approach guarded on both sides by huge stone pillars, each covered with intricately-carved figures. These must have made a great impression on Colin, for, one morning, having reached the trial site in our usual convoy, he suddenly started waving his arms about and loudly, repeatedly, proclaiming that the towers were coming to life, then crashing down upon us. Combined with the rest of the gibberish which he was soon shouting, we could only assume that he was hallucinating, and, whatever the cause, obviously in need of urgent medical attention. We sent him, in the care of one of the other men, straight back to town to seek out a doctor.

All day long we wondered what we'd be told when we returned to our hotel: perhaps he'd suddenly lost his mind, or inadvertently somehow been poisoned, or more likely been bitten by one of the many unpleasant small beasts which all too often plagued us, and suddenly contracted some terrible tropical disease. As it was, we learnt, he'd suffered no more than a 'simple' form of severe heatstroke - it certainly was extremely hot in the Takalar region - and, sedated, was now safely asleep in his bed upstairs. I don't know what the doctor administered, but Colin was back with us the very next morning for our usual Chinese breakfast, highly embarrassed, but apparently none the worse for wear.

Jerry was also finally fit to join us in the field, stomping around the trial plot on his plastered leg without any apparent discomfort. Having spent a while observing the cane growth - not very good, it turned out - the rest of us non-agriculturalists occupied our days by making the usual enquiries about transport and port facilities, and looking around the district for sources of water and a possible factory site. (In the course of this, incidentally, we met some UK civil engineers who were working on a separate nearby project. By complete coincidence, I met one of them again many years later, long after I'd retired, living in the very same small rural village to which I'd then moved. Sometimes, it isn't such a big world at all.)

We soon moved on to the two other sites, both lying to the north of 'UP', in a flatter area which was then largely devoted to growing rice. The nearest - Borotu, if I remember right - didn't engender much interest at all, so we concentrated on the more northerly site, a small village whose name sounded something like Burundi. The problem here was inaccessibility. After one quick visit by the whole group, it was deemed advisable that only two of us at a time should go to work there, spending a day on our investigations, staying overnight, and then travelling back to our hotel at the end of the second. This, however, then raised the question of available on-site accommodation, soon solved by the locals kindly lending us a 'house' consisting of two hammock-type beds slung between the mud-plastered walls of a grass-thatched hut. The washing and toilet facilities, for all requirements, were the same as those used by the villagers: the irrigation ditches. Still, we had boiled water, we'd taken food with us, and at night we had both candles and, more importantly, mosquito coils to burn at our 'bedsides'.

The access issues were simply down to the whole surrounding area being planted with rice paddies, the only means of getting about involving traversing the narrow, raised bunds which ran between them. This was a wholly-pedestrian endeavour, made doubly difficult by having to squeeze past the many paw-paw trees that had been planted at intervals along their tops, and then negotiate around the irrigation gates which allowed water to pass from one field to the next. The main village was set upon the only high patch of ground, and the locals mostly travelled around on bicycles, which even they pushed along these slim, bund-top pathways, only riding when they came to a main, broader 'road'; or, as we would have termed it, a tiny dirt track. Most of my visits, therefore, were devoted to nothing more than walking as far along as many of these raised paths and tracks as I could manage from the nearest point which our Jeep could reach, making it back to our little hut before dark, and then repeating but reversing the process, returning back to the same meeting place by, as if it really made a huge amount of difference, some nominal version of a different route.

The area was, as it turned out, actually quite suitable for development as a sugar estate: but we didn't recommend it to the ministry, as it would have required huge drainage systems, tremendous new infrastructure works, and the relinquishing of a valuable rice-growing site.

It was also the last place to be visited by the team, and so, after returning to our Ujung Pandang hotel for a quick review of overall findings, we said goodbye to endless Chinese food, and the next day returned to Jakarta. Several days had been allocated for us to write up and present an advance report to the minister of agriculture; the final calculations of costs, and our formal viability recommendations, would be completed back in England. For once, therefore, we had a little spare time for some sightseeing and shopping.

It was, by now, nearing December. In that particular year, it was also nearing the end of Ramadan. Indonesia was predominantly a Muslim country, so there were readings from the Koran on the hotel's television every night, and visible preparations were being made for the festival of Eid al-Fitr, when decorations would be lit in the streets. The celebrations, by ancient Islamic custom, could only commence once the senior Mullah actually saw the new moon: but the government, needing to plan for its employees' holidays, had decreed that the official end of Ramadan would be determined, not just as to day but exact time, in accordance with astronomical data derived from modern scientific tables. This divergence of practice was causing huge arguments between the country's more and less traditional Muslims: but the government held fast to its plan. At the appointed time upon the appointed evening, therefore, the Muezzin's call issued forth from the minaret of the city's principle mosque, only a few blocks from, and in clear sight and sound of, my hotel room. The streets' vibrant illuminations were immediately switched on, and there came a sudden, tremendous, trumpeting blast of car horns from the streams of vehicles passing by below, like as not already off on their ways to their traditional parties.

Surprisingly, the street decorations lasted only a few days, and then were hurriedly taken down. In their place, however, instantly sprung up the appropriate symbolic replacements to mark Hanukkah, the Jewish Festival of Lights. Many parties were also apparently held.

And then, not to be outdone - it was a very busy few weeks - it was the Christians' turn, and the illuminations were once more changed over, this time into Santas, stars, and manger scenes. At least this time I did actually get an invite to one of the many traditional St. Nicholas's Day parties, the festivities having been established around that date by the island's early Dutch settlers. It was, like its Muslim and Jewish counterparts, a wonderful demonstration of the Indonesians' shared culture, each welcoming not only friends of their own faith to join them, but those of all other beliefs. And it was certainly a memorable event, the great saint himself eventually arriving, to everyone's great excitement, accompanied by his 'Nubian slave', and then proceeding to blacken the faces of all the strangers present - including me - whatever their natural skin colouring, before shaving them with a large mock razor. With the ceremony thus concluded, presents were given out.

The day of our departure soon arrived. We didn't need to be at the airport for the journey's initial Singapore leg until early evening, and our tickets had already been 'confirmed' with the airline in advance, as was normal practice wherever one was. We were, however, mindful that our onwards flight to London was, by its charter nature, a strictly fixed date, fixed time, fixed destination booking, and unable to be altered, only forfeit, if the connection wasn't made. Still, the 'local run' left us with plenty of time to spare, and the planes usually arrived pretty much on time. We didn't anticipate any particular problems.

Given that we therefore still had a good many hours to kill, and as this was our last chance to buy some souvenirs and gifts for our own forthcoming Christmas, Doug and I decided to make good use of the time, strolling off to explore a large 'department store' some way from the hotel. Having completed our purchases, we popped into the travel agents next door, just to check that there was no last-minute delay to our outbound flight.

"Ah, so sorry," came back the unexpected reply.

Our airline's crews had suddenly decided to go on strike, and it wasn't currently known if and when their services would resume.

Fortunately, the tickets for this first, scheduled part of our trip were transferable. Not so fortunately, and not at all unsurprisingly, news of the strike was quickly filtering out, and all spare seats on alternative airlines were rapidly being snatched up. After some considerable, if hurried, searching, and many frantic phone calls and negotiations, the harried agent eventually managed to secure eight seats on the only available plane back to Singapore which, given the time constraints, remained a viable option. Jerry, because of his late arrival, held an ordinary 'open' return back from Singapore on a scheduled flight, so had intended to travel separately in any case.

Of course, having only checked our flight's status on the off-chance, I didn't have any of the others' tickets with me: but the new airline agreed that, in the circumstances, with the aid of the agent's appropriate hand-written note and receipt, the changeover could be effected at the airport's check-in desk. Once we got there... For this, now, was a serious problem: our new final check-in time was only two hours away, Doug and I were still nearly a mile away from the hotel, and we couldn't even be certain that the other team members would be there in any case. If they weren't, we had no means of contacting or rounding them up. What's more, none of us had yet bothered to finish packing, since we'd anticipated plenty of time in the afternoon.

There wasn't, of course, a bus, taxi, or other form of transportation to be had. So Doug and I simply ran back as fast as we possibly could in the humid, high-noon heat, and, with great relief, if panting severely and soaked through with sweat, found everyone either in their rooms or at least still within the hotel's grounds. They all got busy packing whilst we arranged urgent taxis. Within an hour of first having received the bad news, we were all on our way to the airport.

Happily, for once, given our very last-minute check-in and subsequent dash for the plane, the ticket exchange arrangements went extremely smoothly, and all according to plan. Unhappily for Clay, though, the plane was now fully loaded, and the new airline were utterly unwilling to transfer the previous specially-arranged handling and prepaid extra baggage allowance for his very large, very heavy soil auger. The rest of us, of course, with no time to argue, nor to make alternative arrangements, were more than willing to abandon it, and pretty much anything else, purely in the interests of actually just getting home. Clay, as he made repeatedly clear for the entity of this and our subsequent connecting flight, and for some good time later, I recall, was acting solely under the duress of being left abandoned there himself. It might, for him, even have been a close-run call. He was very fond of that auger, apparently.

Sometime over the last several decades, somewhere behind a thick, solid column in the check-in area of Jakarta airport, someone might well have stood for a while, wondering at the purpose of a very long and chunky metal object, with a sharp-edged screw thread at one end, and a huge T-shaped handlebar at the other.

The rest of us, and all of our possessions, were back in England the next day.

7

THAI TEMPTATIONS

(and just a few troubles)

Everywhere within the former country of Siam, from its overall structure to its buildings, its customs, and its inhabitants' ways, there was a curious, fascinating mix of both the ancient and the very modern. Some of its people led a very sophisticated, Western-style life, complete with rather wayward morals; others maintained a very basic, rural, peasant-style existence which had probably changed very little over the centuries, still farming the same crops as they always had, albeit either with quite up to date tools and equipment, or older, more traditional ones now inventively adapted to suit current requirements.

By 1980, the country was already well-served, and well-covered, by a large, highly-developed sugar industry, complete with its own excellent research station. My initial visit, however, along with Ed Alursin and Victor Knight, had nothing at all to do with it. Instead, they wanted to know if it was possible to expand the commercial growth of cassava. This, in turn, was to serve the establishment of an ethanol-production industry, for which the root crop would provide an additional feedstock, along with old milled cane. It wasn't alcohol for human consumption in the usual sense, though: as increasingly common in Brasil and several other countries which I had or would come to know, it was to be used as a vehicle fuel, substituting for increasingly-expensive imported petrol.

The businessman behind the project was clearly very wealthy, and, amongst other things, owned a chain of high-class hotels, including Bangkok's premier Dusit Thani. It was, in fact, within this very establishment that we, upon our arrival, were accommodated. Which was nice.

He was also, it seemed, an extremely busy man. Even on a pre-arranged visit, we still had to wait a couple of days before he could fit us in for an interview. Accordingly, we made our initial contacts with his office - also in the same hotel - through his personal assistant, a lady not only impressive in her professional duties, but extremely attractive in her appearance, always looking as if she'd just stepped straight out of a high-class fashion magazine. Only on a subsequent, entirely different trip were we informed by a commercial rival - so, of course, he may have had the odd ulterior motivation to bend or exaggerate the truth - that she provided many more services than those of a mere administrative and secretarial nature, with the emphasis most definitely on the personal. But then again, as we also later found out, this did prove to be the general rule, rather than the exception, in most Thai executive circles.

Our client, when we did finally get to meet him, proved to be a rather austere, aloof, and cold character; but he also spoke excellent English. He very quickly arranged for us to be provided with a vehicle, driver, and another of his English-speaking aides: just as competent, but sadly a little lacking in several of the other's more lovely attributes, and also male.

And so, without further ado, we were dispatched off into the countryside. Heading pretty much due north, we arrived by day's end at the little town of Nakhon Sawan, where we were shown into a small but comfortable hotel: and when I say small, I mean that both its restaurant and bar together would have fitted comfortably into any large domestic lounge back home. Having seen us safely to our rooms, both the assistant and driver took themselves off to wherever they were staying, leaving us in the care of the three young ladies who not only ran the hotel, but did the cooking, waitressing, and maid service, too; and again, unfortunately, spoke no English at all. We were able to obtain a drink by the usual gesture system, but once more came the embarrassing problem of ordering a meal, compounded by the florid curls, circles, and flourishes of the menu's script, which, to our ill-educated eyes, looked very pretty, rather resembled - though was obviously not - Arabic, and was otherwise just as meaningless. Still, since out here in rural Thailand we had little hope of knowing what the local dishes actually consisted of anyway, even if described - and, frankly, from some of what we'd at heard, and could certainly imagine, we weren't fully sure that we really wanted to know - we could but adventurously avail ourselves of random choices, and make mental notes of what they were for future reference. Thankfully, they were all perfectly edible. We then just had to work out how to ask at what time we were expected for breakfast, and we were done.

From that time on, we began to establish a form of educational dialogue with the three girls, with us first learning from them the basic numerals, fingers doing the counting, and, by pointing at various things, discovering a few Thai words for the more essential and edible items. The girls, in turn, were very keen to learn the English equivalents. By the end of our two weeks in their hotel, we were all able to have at least some basic sort of conversation, and our stay consequently passed very pleasantly indeed, with the ladies' moral standards at all times remaining nothing but high.

We were collected the next, and each subsequent morning by a driver and our allocated assistant, and thus began our tour of the surrounding area, assessing water resources and examining hectare after hectare of scrubby, but far from under-utilised land, all of which was at least partially, if not always efficiently, planted up with cassava. We ate our lunches at whatever village we found ourselves in, the general arrangement always seeming to entail sitting at a table set outside a residential house where they also helpfully-provided food for passers-by. The meals were prepared on another nearby table, using ingredients which one could select - and mostly identify - from a glass-fronted cupboard, it all then being cooked up in a wok, in turn heated over a stove fired by bottled gas, and served straight from the pan: fresh, clean, wholesome, and thoroughly enjoyable.

Another near-ubiquitous feature struck us as we journeyed around, for almost every village visibly possessed many small to medium-sized workshops, outside of which were displayed some amazing, yet wholly locally-fabricated agricultural tools and machinery. These were often variants on irrigation pumps - much of the land being watered from the multitudinous streams and swamps which dominated the area - usually comprising nothing more than a long pipe, an intake filter at one end, and a pump at the top, their motors being driven by petrol-powered lawnmower engines, apparently imported by the thousands for just such a use. Other common items included two-wheeled rotary tillers, which also had many uses other than just turning over the ground: by detaching the cultivator tools and fitting a towbar, for instance, they became an articulated 'horse', hitched up to a vehicle trailer. Over time, I saw quite large and heavy loads being pulled by such units; not to mention their inevitable ancillary usage as a means of public transport.

We returned to Bangkok at the end of our allotted time, wrote out our report, and then, having had his delightful secretary type it up for us, presented our findings and recommendations to the great man in person.

I don't know what became of the project, though. It certainly wasn't something with which I, nor any known colleague, was ever involved again.

Come 1982, and it was back to normal. The Mitr Phol group controlled a large number of industrial enterprises, including sugar companies, and they currently wanted a new, all-in-one cane processing plant, capable of crushing a sizeable 15,000 tons per day and then refining it into white sugar. Our consultancy company's engineers were, of course, more than happy to oblige.

Alongside this, but far from insignificant in its own right, the agricultural team was asked to survey an area of land extending for many miles around the proposed factory site, to determine where more cane could be grown, and then both encourage and assist the local farmers to plant it, also providing them with whatever initial advice, requisite equipment provisioning, and general ad-hoc technical follow-up assistance they might require in order to establish and subsequently harvest the crop. A fairly large part of this programme ultimately fell to me, both in terms of supervising the agricultural side of the contract, and also in designing a harvesting and handling system, together with all of its equipment, which transported the cane all the way from the fields right up to the point where it was placed upon the conveyor leading into the mill.

We initially sent a team of about a dozen design engineers out to Bangkok. Housed in the Bangkok Palace hotel, everyone, with the aid of a dedicated minibus, was then shuttled every morning and evening to and from a suite of private offices in a city-centre building owned by Mitr Phol. The company was nothing but generous, and their president, Mr Saetang, and his deputy, Mr Vithrong, were always extremely friendly towards us, keen to ensure that we were all provided with every possible comfort which might be needed; well, excepting that one, of course. Though it probably would have been made available, had it been requested: and, for all that I know, it even was. Whatever my colleagues got up to in their non-working time was - as far as I could possibly help it - none of my business, and I shan't comment further upon that which I really didn't wish to know.

Our offices were located at the edge of Patpong district, a busy commercial centre housing many a corporate head office. Its streets, as with everywhere else in the city, were also extremely busy, filled with 'tuk tuks', those ubiquitous motorised rickshaw-style tricycles - two wheels at the back, this time - fitted with a motorcycle engine and a covered rear bench seat for three people. Colourful, sometimes makeshift, and occasionally pretty rickety, all of us nevertheless found this plentiful mode of transport very useful for getting about in our 'out of work' time, and much preferred them over the standard taxis, for not only were they much cheaper, but they could weave their way through the alleys, markets, and smaller back roads, and thus avoid the congested main city routes.

We didn't need to venture too far from our workplace in the day, though, as there were several fine restaurants within the district, and, just a few short minutes' walk away, an 'English pub'. Here, if so wished, one could not only drink imported 'proper' British beers, but also eat 'proper' English dishes such as steak and kidney pie, roast beef, and sausage and mash. The atmosphere was good, too, so we often enjoyed our lunches there, albeit always drinking the local brew.

But we never went there after dark. Nor, with the exception of the excellent Steamboat fish restaurant, did we ever brave spending our evenings anywhere else within the district. For Patpong, at night, was also the centre of the Thai porn trade, its vibrant, garish streets then thronged with tourists of all shapes, sizes, types, and nationalities. Some were simply running the astounding gauntlet of its persistent touts and crushing crowds in search of a decent meal. Most, however, were on the hunt for satisfying refreshment of a rather different kind, amply catered for by the many and various shows, parlours, and outright plain prostitutes. Or maybe both: dinner and afters, or entrees and main course, dependent upon your personal preferences and priorities.

To me, however, one of the best, and certainly most amazing sights didn't have anything at all to do these particular night-time attractions. Well, maybe strictly not true, for I'm told that the area's combined day and night-time businesses' overall needs for reliable and instant power and communications, together with the glaring, multicoloured lights and signs, and all of the buildings' and restaurants' exhaust and ventilation systems, provided an excellent warm and spacious roosting stop for the hoards of swallows which passed through on their annual migration. Every September, come dusk, every overhead power and telephone cable would become lined with the birds, thousands to a street, all perched side by side and spaced a precise wing's distance apart; and, accordingly, beneath each lay a neat, growing row of almost mathematically-placed white spots. Great care, skill, and judgement was required whenever you had to pass under overhead wires.

Surprisingly, with all of the above notwithstanding, I have to say that Bangkok was a very easy city to walk around without being pestered or feeling at all ill-at-ease; in the daytime, at any rate. Nor was it ever unpleasantly hot. As such, whenever I did have some time to spend there, including some later weekend visits, I did manage to see many of its other famous attractions, including the Palace, the Golden and Reclining Buddhas, and the magnificent Royal Barges; not to mention the very many markets where one could buy an amazingly-large selection of 'guaranteed genuine' designer clothing, and a multitude of other similarly-authentic goods, all at extraordinarily cheap prices.

It was agreed that the design team would stay in Thailand for two years, firstly producing the detailed engineering and construction drawings for the new factory's structure and plant, then moving up to site to supervise the erection and installation of the buildings and machinery, and finally overseeing the start-up process itself. I was scheduled to make frequent visits out from the UK, each of around two months' duration, working alongside two resident site team members: a civil engineer who supervised the irrigation and drainage aspects, and Phil David, an agriculturalist who also possessed considerable experience in mechanical cultivation.

The factory was to be built in the central north-east of the country, between Bangkok and the Laos border, on the outskirts of a small settlement called Chaiyaphum. As was so often the case, however, the nearest suitable accommodation, even on a temporary basis, was about seventy-five miles to the north-east, in Khon Kaen. Here, not only were there several reasonable hotels, but a university, a teaching hospital, and an airport serving many flights a day to and from other parts of the country; so here, therefore, we set up our operational base. In due course, Mitr Phol found us some space in a shared house near to the new factory, so we at least then had a permanent site office. The remainder of that property, however, was occupied by a group of Thai surveyors, and whilst there was absolutely no conflict as far as general matters were concerned, every lunchtime they took over the kitchen en masse and began to fry garlic, its particularly-pungent smoke quickly penetrating every single room. This, to our sensitive English noses, was the signal for evacuation: and so, in the middle of each and every working day, we downed pens and paper, picked up our pre-prepared packed lunches, walked well away from the house, and sought a few hours' alternative occupation until the air had cleared again. Never its biggest fan in any case, I can't even now smell cooked garlic without thinking of Chaiyaphum.

But there was still the long daily drive, its monotonous entirety through open, featureless country, broken only by one small halfway town and a regular police checkpoint. To while away the time, Phil purchased several packs of cassette tapes, each featuring recordings from well-known BBC comedy shows, such as The Goons. Even spread between our two pick-ups, we soon got to know all of the scripts off by heart, as they were played over and over and over, every single blinking morning; and then, all the way back as well. Yet despite their questionable Bangkok provenance, they never wore out, and they never broke or even needed rewinding onto their spools with that great technological aid, the pencil.

I'm quite surprised that, one day, someone didn't just throw them out of a window.

At least my sanity got some respite, as I still had to spend a considerable amount of time in - and therefore back and forth to - Bangkok, meeting frequently with both the Mitr Phol management and our factory designers. The majority of my off-duty time, however, belonged to Khon Kaen, where, through our contacts at the university, we soon made a number of friends amongst the local population, including members of an English-speaking church which many of us visited several times.

One bored Sunday, Phil and I decided to take the hour or so's drive north to where the Mekong river formed the border with the country of Laos; just to see what there was to see. It wasn't much, as it happened. There was supposedly a ferry service at Nong Khai, and we did indeed see one single, solitary boat, constantly travelling to and fro along the river's centre line: but it never actually made a full crossing. (We later learned that this stretch of the river was strictly patrolled by the Laotian police, lest any of their citizens ever tried to escape over and into Thailand, and that the ferry, for this reason as much as any, no longer ran.) After coffee at a local bar, we drove for a while longer out along the riverside road, hoping that there might be something else of interest to see or do. And then we noticed a large number of what appeared to be tall statues, standing in a nearby field. We stopped, and had a wander around what turned out to be the most amazing collection of highly-coloured, if extremely-grotesque, carved and painted figures, some standing at over twenty feet high. They looked to us rather like some of the Indian deities, only a good deal more Satanic.

Quite excited about our find, which certainly wasn't mentioned in the list of nearby attractions as supplied by our hotel, we were soon telling our local friends back at Khon Kaen all about it. Their expressions immediately turned to pure horror, as they warned us in no uncertain terms never, ever, to go there again. The site was a temple of sorts for some Voodoo cult, they explained, and accordingly was a most dangerous - in all kinds of utterly non-specified ways - place for non-members to be.

We never seemed to suffer any harm from our unintended trespass upon their shrine; but we never paid a return visit, either.

Being full-time residents, we also became quite amicable with the staff at our hotel. Like so many similar establishments, the restaurant had a resident band and singers: and they, along with the waitresses, used to sit and talk with us in the lounge during their off duty times, making the most of their novel guests' presence to improve their language skills. Again, though, it was a strictly social and platonic arrangement, for our hotel made no concessions whatsoever to their country's best-known industry, and taking girls (or anyone else) upstairs was, they made clear, most strictly prohibited. Apart from the lift doors being in the full sight of the reception desk, there was always, at the top of each floor's stairway, right next to the lift, an employee whose sole duty it apparently was to make ensure that no 'hanky-panky' took place within their realm. Of course, bribery, was even more rife than the sex trade, so I suspected that these 'floor managers' could quite easily have been persuaded not to have been looking in the appropriate direction, had one ever wished to avail oneself of the local 'talent'.

Not that I ever put it to the test.

The hotel's attitude definitely wasn't replicated elsewhere throughout the town. In addition to their dining areas, most restaurants also had a dance floor, the standard layout usually including a large, brilliantly-lit 'display case', behind whose long glass wall, lined up on tiered rows of clearly-numbered chairs, sat a number of potential female dance partners. All were young and attractive, and all were available for immediate hire, should one not already have brought one's own. Every so often, therefore, someone would be seen signalling a waitress, and asking, by number, for a girl to be called. She would then promptly come to the table, possibly have a drink with the diner, and then dance with them for as long as was required; and, presumably, paid for. Procedures thereafter seemed to vary. In many such establishments, the young lady simply returned to her seat to await re-selection. In one place where we dined frequently, however, there was certainly one very notable door through which, after a dance or two, the young lady and her gallant escort would often be seen to disappear, returning only some considerable time later. We couldn't possibly guess why.

It was a pretty similar story at every place of entertainment, whether a theatre, floor show, or the bowling alley. These young female 'partners' - and they were always, at least as far as we could tell, female - frequently appeared, wholly unsummoned, as soon as - sometimes even before - one had taken a seat or table, the attending number varying only according to the count of unaccompanied males. The fact that we spoke only English was usually sufficient to put them off: but sometimes, one or two would persist in trying to chat with us. It wasn't unpleasant, and nor did either we or they get uptight about it. Mostly, we just found it vaguely amusing. According to the inexplicable howls of laughter which they sometimes produced during our odd, innocently-intended interactions, apparently so did they.

Still, it could sometimes get quite embarrassing: if not always in the expected ways. One evening, Mr Saetang and Mr Vithrong, along with another of Mitr Phol's senior managers, came up to Khon Kaen. Straight away, they insisted that we - a sizeable party in terms of numbers - joined them later at a floor show. Immediately upon our subsequent arrival, the Thais ordered up the requisite number of female companions, distributing them amongst us on a one to one basis. Drinks were then obtained - priorities obviously had to be attended to in the right order - and everything seemed fine: until after the show had finished, the lights were once more brightened, and we realised that all three executives, along with their girls, had vanished.

We waited around for a while, and then, realising that we'd now been left to our own devices, promptly waved goodbye to our own young ladies, who couldn't speak a word of English anyway, and grabbed some taxis back to our hotel.

The next morning, our unabashed clients turned up for their scheduled meeting and quite casually enquired if we'd enjoyed the night. They were very surprised to learn how it had ended, explaining that the cost of the girls' hire was wholly and gratuitously on their account, and had covered the full period until daybreak. Fortunately, they didn't seem to be insulted: but it was a considerably-awkward moment. They simply couldn't comprehend why we hadn't seemingly understood, never mind hadn't availed ourselves of, the total entertainment package on offer. It was, after all, nothing but that which any self-respecting Thai gentleman, of whatever marital status, would have done.

The site's staff accommodations began to be built at the same time as the factory, so my long-suffering commuting colleagues, together with those who'd by then come up from Bangkok to supervise the factory's construction, were finally able to move into a sizeable guest house, which I, upon my subsequent visits, also shared. Moreover - for which my garlic-assaulted nose was forever grateful - the field advisers and I were then assigned the entirety of a newly-built office facility, destined eventually to be Mitr Phol's agricultural department, once everything was fully set up. We were also allocated the secretarial services of two local Thais, both young men, one of whom was fluent in English. He'd been schooled as a Buddhist monk, as were many of the higher-educated population, and turned out to be an extremely interesting person to talk with, both during and outside of working hours.

By now, I'd drawn up specifications and made the appropriate technical drawings for the transport and cane-handling equipment which I considered to be required. The Bangkok-based board of directors had approved them readily, adding that they'd have everything made up locally to my exact designs. Of course, given that the country's whole economic system at that time worked on bribery, corruption, and barter - as did so many others, to be fair - I shouldn't have been so surprised to find that the trucks, when finally delivered to site, differed from what I'd originally intended in several important ways. Upon enquiry as to the reasons behind the unexpected changes, I was told that the management had simply purchased what the vehicle's makers considered to be the nearest existing models to my drawings. This, it was explained, meant that no unnecessary, excessive, or unusual costs would be incurred. What they really meant was that someone had done some sort of deal, and never mind the consequences: but the latter, I knew with a sinking but experienced heart, sooner or later, to some extent or other, there would certainly be. Accompanied, quite probably, by a lot more expense.

Our design engineers had also been told that, apart from the mills themselves, all plant, machinery, steelwork, and allied buildings for the factory's construction would be locally manufactured. This, we'd all assumed, would mean that it would be made in the plentiful workshops; and, hopefully, somewhere nearer to design -

and designers' - tolerances, too. But then one day, to the surprise of everyone working near to the actual site, ton after ton of raw steel - box sections, girders, and plate - suddenly started being delivered to, and piled up within, a compound which was scheduled to become the cane yard. Next came a number of large machines, including plate-bending rollers, drilling and turning plant, and a very large horizontal borer. Add to this a multiplicity of cutting, sawing, and welding tools, and it soon became patently clear that 'local manufacture' meant exactly that.

None of us had ever seen a factory and its constituent components actually being assembled right alongside their final location. This time, however, we were able to watch and assess daily progress right before our amazed eyes, as the framework was sized, shaped, secured, and erected, and the huge vats and vessels formed from scratch, then placed directly into their final positions. It certainly had something going for it in terms of efficiency.

We also watched the unconcerned construction workers as they walked along the sickeningly-narrow beams, and shinned up vertical metal columns, to reach their work stations, with only a bare minimum of scaffolding and some extremely-precarious platforms and partly-built flooring beneath them. That, we didn't so much care for.

Just before the factory's completion, I was asked to prepare a formal induction and instruction programme for the operators of the harvesting, handling, transportation, and cane-yard equipment, and also to conduct some initial in-person sessions prior to the official start-up. These tasks, as such, were not a problem. Most of the designated men, however, along with their associated supervisory staff, not only weren't at all familiar with the systems being installed, but wouldn't know any English. I therefore first had to train up my monk-cum-secretary-cum-translator in all aspects of the relevant technical matters: and that was no mean feat, since he'd never even seen a single part of the sugar cane growth, harvesting, and processing cycle in person before, never mind been inside an actual mill. But he was a very quick and able learner, and, once fully cognisant of all the theoretical issues, soon had all of my hand-written scripts converted into Thai, then printed up.

Embarking on my last scheduled visit, therefore, there was very little else left to do. A few outstanding queries to be rounded off, some final advice to be imparted to those working on cultivation methods in the fields - to which end I also brought out Vic Knight, last encountered on that first Chaiyaphum assessment - and the training sessions to be held for the more senior staff.

All seemed to go well. My eager and willing students all appeared at the appointed location on the appointed day, even in advance of the appointed time. The translated notes were handed round, and, once more with the aid of my now-knowledgeable secretary, my every sage pearl of wisdom was imparted, and only occasionally questioned. By the end of the week, the attendees all felt not only confident of their future tasks, but ready to pass on their new skills to the general workers. Or so, anyway, they apparently said.

And my work there was done.

I did actually have one more brief visit to pay before formally departing for home, flying over to a general crop and cane estate at Kamphaeng Phet, about two hundred and fifty miles west. I'd agreed to loan Phil David to its owner for a short while, and I wanted to check that he, and generally everything, was all right before I left. Phil, I found, was not only faring well, but was positively living in the lap of luxury, having been assigned the most sumptuous hotel accommodation, and generally being treated as if his was the first, last, and only word in agricultural advice. Still, as I was the one who'd initially arranged his services, I, too, also benefited from some overdue thanks, and was very well-received and treated throughout my short stay.

And so, with the satisfaction of several jobs well done, I went back to Chaiyaphum, picked up Victor, bade my farewells, and flew back to Bangkok. We made our respective reports to the Mitr

Phol directors, and then prepared to end what was to be my last ever overseas visit as a paid, professional consultant. My 65th birthday was in less than a month, and I was due to retire.

But first, there was Christmas. To be precise, there was a good two weeks and several days before the celebrations in my home country, and I'd booked return flights secure in the hard-earned knowledge that there was still plenty of time left to get back, even should anything or anyone as yet unknown still dare, this one last time, to try and stand in my way. But there was nothing now to prevent me from leaving. Nothing else to do, nor to be done, except to relax, while away our last few Bangkok nights, and spend our remaining days shopping.

Oh, poor deluded fool!

It was during our first free day's foray that Vic and I dropped into the rather contrarily-named Office of Immigration to obtain the requisite tax exemptions and departure visas: a mere time-consuming formality which we'd dutifully undertaken upon each past visit to the country - not to mention the many, many others which I'd entered and left - and with whose ropes, as old dab hands, we were now thoroughly and wearily familiar.

It was the usual Thai set up: row upon row of little desks, each complete with an officious-looking clerk, each with but one single operation to perform. The practice was to start with the enquiry desk nearest to the door, and then to progress through each link in the chain, one man and one desk to the next, until all procedures had been dealt with, all details and pitfalls successfully negotiated, all forms obtained, filled-in, handed over, examined, checked, initialled, rechecked, recorded, filed, reissued, re-stamped, added to, signed, stamped again, and so on; and, at the very last, collated into an official clearance paper, along with a firmly-embossed passport page.

This time, however, we never even made it past the first clerk.

Having handed over our completed application forms and passports, the official, as he always did, promptly scurried off through a door to consult with - or, at least, to make some show of consulting with - his superior.

And there he then stayed.

We quickly began to suspect that something wasn't quite going to plan. Sure enough, after a lengthy delay, the clerk reappeared. He asked us to accompany him into a side office.

We'd only ever encountered desks and clerks before. Side offices didn't sound good. Senior clerks didn't sound too inspiring, either: and the one who sat in front of us - at his own separate desk, in his very own separate office - didn't lend any more confidence, immediately beginning to question us in depth about our work in Thailand. Or, rather, he began to try: only he spoke very little English, and we, of course, had no Thai at all. Translator clerks didn't seem to exist, and we hadn't thought to need the services of our own. Then again, we'd also thought that we'd filled in the tax forms exactly as we'd done every single time before, never with even a hint of an issue, so we really couldn't understand what the problem was now.

It all proved to be a rather difficult conversation, the impasse only finally being resolved when the official decided to take us through to another office, this time containing an even more senior colleague, who could at least speak a little English.

He also, worryingly, had a bigger desk.

The problem was, this next man patiently informed us, that no-one could understand how or why it was that we'd been working in Thailand for so many months, never mind on so many preceding occasions, and yet hadn't apparently earned any money at all.

We, by return, attempted to explain as best we could - and not for the first time, so it wasn't exactly by now a novel, nor unpractised spiel - that we were travelling international consultants, and that, in common with all kinds of other people doing similar kinds of jobs, our fees and salaries, less any UK taxes required to be deducted, were paid to us only in the UK, our own home country, the place both of our permanent residence and permanent employment, by our UK-based employer, for whom we worked both there and around the world, and for whom, and on whose behalf, and in whose pay, we had been doing our temporary work in Thailand.

But it was all to no avail; and, seemingly, all beyond his comprehension. If, he argued, we were physically working, giving advice and providing training in his country, then we must, prima facie, be earning a salary there.

We countered by stating that our earnings were apparently regarded as 'offshore' under his country's tax regulations, not to mention various international agreements - or so we'd reliably been informed by all kinds of lawyers and accountants, both in the UK and Thailand; all arrangements, we'd been repeatedly assured, perfectly legally-acceptable - in accordance with which we'd already been previously and repeatedly certified and cleared, many, many times, by this, his very own department.

He was still having none of it. Moreover, he refused to telephone Mitr Phol, or anyone else in the UK or Thailand, for verification. Nothing, he said, could or would be done until we returned to present a formal statement of our earnings, thus properly to be assessed for whatever taxes we must pay. Only then, and once all of said dues had been settled in full, would we be allowed to leave the country. And until such time, furthermore, lest we should manage to find another way to sneak out without paying - over that river into Laos, perhaps? - our passports would also be impounded.

We were then, rather brusquely, shown the door.

Vic and I, needless to say, weren't at all happy with this turn of events. It wasn't, of course, my first brush with this kind of problem: but it also wasn't one which I'd ever anticipated, nor previously encountered even the slightest whiff of, in what I'd until then considered to be quite an enlightened, educated, and efficient country when it came to such matters. I could only therefore conclude that this was some odd attempt on someone's part to extract some form of backhanded bribe.

Or perhaps the payback from those Voodoo statues..?

We immediately hot-footed it over to Mitr Phol's offices, only to find that all of the senior directors were out of town, and that nothing, terribly sorry, could be done until their return. So back we went first thing the following morning, and anxiously explained our problem to Mr Vithrong. He, promising forthwith to investigate, could only ask us to contact him again the next day; and the next, and the next.

Our carefree, pre-Christmas cheer had by now turned to desolate gloom: but there was nothing which we could do except to bite our nails and keep hoping. Eventually, on the twentieth of December, we were summarily officially informed that the matter was finally settled. Not even stopping to ask what, how or why, Vic and I rushed straight back to the immigration office to collect our passports and, just as importantly, those precious tax clearance papers, before someone again changed their minds about letting us go.

It was Mitr Phol, we later learnt, who'd finally agreed to foot the considerable bill for our release. Exactly what the argument had been, and who had actually been right or wrong at what stage in the game, we never cared to fathom. We knew only that someone, somewhere, quite high up in authority, had suddenly decided that they wanted to squeeze some money out of T&L UK in respect of their Thai contract. Holding a couple of its employees as hostaged pawns, they'd reasoned, would probably do the trick.

One can't, at least, say that it wasn't an efficient means of enforcing tax compliance.

By now, of course, we'd well and truly missed our flight home. Hastening straight to the nearest travel agency, we were apologetically, if not really unexpectedly, informed that everything back to the UK - to anywhere at all in Europe, in fact - was, given the time of year, already pretty fully booked.

But not entirely, we insisted. They'd not said entirely. There must be some possible opportunity, we demanded, pleaded, begged, asked terribly, terribly nicely indeed. There must be some means, some thing, some way or other, to get us somewhere at least close...

In the end, I managed to grab the very last seat on that evening's plane out to Copenhagen, with a slim hope of catching another to Heathrow, providing that everything flew to time, and that I could run very, very fast in-between. Vic was similarly able to leave the same night, travelling by a different, and even more circuitously-demanding route.

Of course, having checked-in at the airport with very little time to spare, this, out of all of my departures from Thailand, just had to be the one time when the customs official stationed at the baggage checkpoint chose, instead of the usual cursory glance and verification of the case's ownership and destination, to rifle through my entire luggage with the finest of toothcombs before its placement onto the final loading trolley. Just in case my harried, anxious state meant that I was exporting illicit Thai antiquities, or some such like, I suppose.

Or Voodoo artefacts, perhaps?

Suffice it to say that, albeit significantly stressed, I did make both that flight and my next day's connection in Denmark, and arrived home, surprisingly complete with all possessions, just in time for the full family festivities.

I swore that I'd never again fly, for any reason, official or otherwise, so near to Christmas. December at all, in fact, was generally looking decidedly dodgy. And, indeed, I never did.

8

NIGERIAN NIGGLES

Africa's vast continent is comprised of many countries, all varying widely not only as to their climates and physical characteristics, but with very differing cultures and customs, too. There are some of which I became very fond. Several, alas, didn't exactly endear themselves sufficiently to rank amongst my more favoured parts of the world.

My very first contact with this huge region was at its centre, in Zambia: but that was a comparatively-fleeting visit, during which I glimpsed only the smallest part. My next, much longer trip was to the west; to Nigeria. And there I then went again and again, sufficiently so that, out of all the other African countries, bar one, it merits a sizeable chapter all of its own.

I think that someone was trying to tell me something right from the start, 1971's outwards journey taking significantly longer than that very first 1954 flight. I was travelling out from the UK to Kano, Nigeria's second-largest city, right up in the country's north. From there, I'd then take a local flight down to Yola, over on its eastern border, and just a few hours drive from the site where we were, as usual, investigating the possible development of a large cane estate and processing plant.

Everything went to plan at first: I was collected from my Surrey home by the usual friendly but respectful, grey-uniformed and peak-capped chauffeur - nothing special in those days, just all part of the normal pre-booked local airport taxi experience - and driven over to Heathrow in time to catch my evening departure. Things proceeded smoothly through the check-in and other airport procedures, too, and I duly boarded my Nigeria Airways Boeing 707 fifteen minutes before its scheduled take-off slot. The passengers all sat buckled up, expectantly awaiting doors to be closed and engines started.

But instead, we just sat there; and then we sat there a bit more. Time went by, nothing much of any interest happened, many numb buttocks began to be shuffled, and still we continued to sit: until, at 11.30pm, we were finally informed that an engine fault prevented the plane from leaving that night, and we should all return to the lounge.

The whole of the airport was silent now, every other flight having long since left. Its shops and cafes were closed up for the night, its employees all returned to their homes. There were still a few British Airways staff, attending to the last of their end-of-shift duties, but none from Nigeria Airways. No-one, least of all the passengers, seemed to have a clue what was happening next. Eventually, after a long wait, it was a BA stewardess who came to tell us that our baggage was being unloaded, would soon be returned to us, and that a coach was waiting to take everyone to a hotel, from which we'd be collected again at 7pm on the following day: or, rather, the same day, as it was by now 1am in the morning.

So off we went into the night, about fifty of us in total. But instead of heading towards one of the nearby hotels, the coach turned several times, and in a short while crossed the River Thames. We were, I realised, now heading south on the bus route between Heathrow and Gatwick, a route which would pass within just a few yards of the end of the road on which I lived: and, surely enough, that's exactly what happened, before we were finally deposited outside Gatwick's Post House Hotel.

This was fine as far as I and the non-Nigerian passengers were concerned, and we quickly went off to our assigned beds. For the large contingent of northern Nigerians, however, there was still a very big problem, for they were Muslims, and it was Ramadan. Having not properly eaten since daybreak many hours earlier, they'd obviously expected to break their fast as soon as they were in the air, and so were now, understandably, demanding food. This was not, however, the age of twenty-four hour guest service, never mind supermarkets, and the hotel's catering staff had vacated the premises many hours ago. In the end, I later learned, the rising uproar of so many hungry and increasingly-unhappy passengers was only quietened once the hotel's skeleton night reception-cum-portage staff were hurriedly redeployed to search its cold, empty kitchens for anything which met not only the even-more-problematically-acceptable ingredients criteria, but would also fit into a sandwich.

The next morning, after a hearty, and most definitely non-Halal hotel breakfast, I telephoned home to surprise Eleanor, who'd by then expected me to be far, far away, instead of only a mere few miles down the road. With a logic which must simply have bypassed my tired brain the night before, she politely enquired as to why I hadn't just asked to be dropped off at the end of the road, telephoned the airline the next day to check on progress, and then got another taxi back to the airport the following evening. With no excuse except that I hadn't wanted to wake her, I did indeed then catch the bus from outside the hotel, and within twenty minutes was spending the rest of my day back at home.

Returning to the hotel in good time for our scheduled pick-up, I again checked-in at Heathrow, now joined by another half plane load of passengers arriving for that night's flight. Again, I boarded the plane; and again, after an interminably-long, boring, and decidedly-uncomfortable wait, I found myself, together with all of my fellow would-be travellers, returned to the terminal. Either the airline's engineers had failed to correct the original fault, or something else had now occurred on start up. Either way, the result was the same.

This time, unsurprisingly, many of the passengers became immediately angry, taking their frustrations out on the helpless BA ground staff who were once more trying to deal with the situation in the equally-repeated absence of the operator's own employees. It also took even longer to find a hotel - a good way to the airport's north, this time - and it now required two coaches to convey everyone there, a task finally accomplished at gone three in the morning. The mounting, anxious clamour from the starving Muslims, never mind the apparent inability of anyone from the hotel to understand why they couldn't just wait the few remaining hours until breakfast to eat, was only compounded by the news of there being insufficient single rooms available to accommodate everybody there. I glanced around me quickly, selected someone who I hoped might make a reasonable short-term companion, agreed with him to share a double room, and went off to get some sleep before a full-scale riot began.

Come breakfast-time, we were told that we now had to hurry back to the airport for a midday departure. So off everyone once more went - third time lucky, hopefully - only to find ourselves boarding not the expected Nigerian passenger jet, but an obviously swiftly-converted Aer Lingus cargo place; minus a few seat-belts, for sure, but at least actually managing a successful take-off, so no-one was going to complain too much.

It was evening, local time, when we landed at Kano. It took a good while to track down one of the airline's representatives - they did seem to be terribly good at hiding, even at their own home base - but I finally managed it, explaining that I'd originally been booked to fly on to Yola two, if not by now three days ago.

"Oh, no problem, Sir," came the reply to my initially-relieved ears. "All local flights have been cancelled for the past several days."

This particular delay, I was then informed, was due to the stubborn decision of a dense cloud of fine dust particles, blown down from the northern deserts on an unusually-early Harmattan-type wind, to remain lying over the whole of upper Nigeria, thus not only severely reducing visibility, but in places covering the ground to a depth of many feet.

"We will book you into the airport hotel."

So there I sat, and stayed, checking in each day to enquire if and when they could safely fly. Of course, I realised, even once onwards progress could finally be achieved, I now had absolutely no idea where my field team might by then be - nor, indeed, already were - never mind how I got to them; and, in the meantime, no way to contact them to explain my current predicament. Quite likely, they wouldn't even know of the reasons behind my continuing delayed arrival. Perhaps they'd already assumed that I'd for some reason cancelled my visit, or got lost or taken ill on the way over, or... Well, anything.

On the fifth day after finally departing the UK, I was told that the dust had apparently lifted "to some degree", and the airline were therefore going to attempt - not a thoroughly reassuring word, I thought - a flight to Yola the next day, where the pilot would then try - I didn't particularly like that word, either - to land.

The way that the trip had so far gone, I wasn't feeling entirely confident as we taxied out towards the runway. But the plane at least managed to get as far as lifting its wheels off the ground - which, in Nigerian terms, had to be a sign of significant progress - and then spent the next uneventful hour flying smoothly above what looked like a layer of very thick fog.

And then the plane descended into it: indeed thick, and very, very fog-like.

We were only just above ground level when the visibility cleared enough for me to be able to make out what might, somewhere through the remaining haze, qualify as an airstrip. I assume that the pilot saw it at about the same time, too: and, since this was neither the time nor the type of aircraft where anything but the pilot's eyes, and presumably a radio, were standard equipment fittings, I could only trust that he held the advantage of having been here at least once before, and therefore possessed some basic idea as to how to approach and land upon it. My engineer's brain also tried very hard not to consider the effects of all that swirling sand upon the turbines, lubricants, and whatever other moving parts generally make planes go, stop, and do whatever else they all have to do in order to get me safely from one place to another.

As it was, albeit with the aid of a few sudden adjustments, both the plane and its suitably-skilled pilot performed their required tasks quite adequately, and we few fearless - or foolhardy - trailblazers on board were soon alighting back onto terra firma to reclaim our baggage.

I had finally arrived at my destination. And I had no possible clue as to how, or to where, I should proceed next.

There wasn't a taxi or bus to be seen as I walked out of the terminal hut; just a few parked-up motorcycles, and one private car. I was just wondering about asking for help from its driver, currently slouched, perhaps asleep, inside: and then he suddenly sat bolt upright, opened its door, got out, and began to approach me, repeating my name in a questioning voice.

Was I ever pleased to see him!

I'd bet that he was pretty glad to see me, too. It turns out that the poor chap had been waiting patiently outside this makeshift airport for four straight days, with no information forthcoming from anyone else who was apparently any the wiser, simply hoping for any plane to come in upon which this unknown white visitor might possibly have flown.

My conscientious escort promptly drove me to the southern banks of the nearby Benue river, where a ferry took us, along with scores of others, to the faraway side.

Yola, in the local language, apparently means the land of vast plains, and, after an hour's or so's drive up a good road, we still had the same, suitably-large, subtly-rolling one to our right hand side. To our left, however, now rose an imposing, conical mountain - locally called something like Konan Kuka, I later found - after which a small village straddled the road. My driver soon turned down a dirt track, and stopped about a mile later in a clearing which contained two small bungalows of obviously European design, with an old, but good-sized caravan parked next door. It also contained three surprised-looking T&L colleagues.

Travel delays notwithstanding, I was already joining the site team late, having first had to complete another contract elsewhere. The men who greeted me - Josh Waters, a factory designer, Ned Fender, an agronomist, and once again Jerry Sterling - had therefore already been settled in for a good many days, alongside another six men from a mixture of other employers, plus the project's overall co-ordinating director, Rupert St John. Rupert headed up agricultural operations for the Commonwealth Development Corporation, a British governmental financing organisation formed to invest in overseas agricultural development projects - as it was in this case, partnering with the Nigerian government - operating originally within the 'colonies', then expanding to cover the Commonwealth in general, and then soon just about everywhere else as well. And he was, clearly, a very busy and important man: he'd also had to bring his own assistant out with him.

Expansion was something desperately needed for these cramped accommodations, too, all twelve of us, plus a pair of cook and laundry boys, crammed into a couple of tiny houses previously constructed for the managers of a now-defunct cotton plantation. Upon arrival, apparently, the CDC pair had promptly decided that they should each singly occupy the two available bedrooms. The original T&L trio had therefore swiftly commandeered the four-berth caravan - a move which turned out to be the best, and definitely quietest, option by far - leaving the remaining six to make do with hammocks slung across one bungalow's sole living room. The other's similar space had already been turned into a communal dining area, as well as a sleeping area for our 'houseboys'. It also accommodated two large, generator-powered freezers, one containing a good stock of food, the other chock-full of what CDC clearly considered to be their most essential field stores: bottles of high-quality wine.

The team had already set up a daily programme, although so far only exploring the more easily-accessible areas. Some members went off individually each day to study water resources, soils, and so on, leaving the four T&L men, together with Rupert's trusty assistant, Stuart Waldron, to share the two remaining Land Rovers between them. The Great Chief, naturally, stayed safely 'at home', writing up his notes and collating reports from the daily verbal observations of those bravely venturing out into the wilds.

Ned had already done much of the initial visual surveying, having first been seconded out here over a year ago, so proved to be an invaluable guide. He'd also, in the course of his work, cut out a trace (or track) west to east through the Numan Plain, where it was hoped that the cane would be grown. Early one morning, therefore, well-supplied with packed lunches and several bottles of water, our mini-convoy set off to travel the full length of the 'Fender Trace'.

The plain was covered with dense grasses, standing at least six feet tall. Ned's thin track was wholly indistinguishable from a seated position, so we all took turns riding on the roof of the lead vehicle and directing our local drivers. It was slow going, and, apart from stopping now and again to examine the soils, nothing of any great use was gained. At lunchtime, having pretty much reached the other side of the plain, we all agreed that it was pointless to return the same way. Instead, we decided to head south for a couple of miles, travelling through the rough, uncut grasslands on a compass bearing, then turn back on a parallel to our outward route.

All went well at first, even if we did have a puncture: but it didn't take long to change to a spare wheel, and by mid-afternoon we reached the end of our southward leg.

And that's when the trouble started. We hadn't gone much further when we drove unexpectedly out of the high grasses and straight into something which could only be described as a gigantic dry swamp. In fact, had it been the wet season, it would undoubtedly have been a giant and very wet swamp: a classic 'gilgai', as it was technically known. Everywhere around us, thick tussocks of reed-like stalks sprouted up to form endless small mounds, each distinctly separate, yet close enough that one could easily step from one to another. Still, although the going now became significantly more uncomfortable, our trusty Land Rovers continued to make slow progress, bouncing along from one raised clump to the next with only an occasional sunken wheel in-between. The further that we travelled, however, the higher and larger the hummocks became, until the wheels at times couldn't reach the ground between them, the vehicles became stuck on their sumps, suspended in mid air, and we all had to disembark and undertake the soon-repeated, soon-continual, and certainly laborious and time-consuming task of knocking down each dry, hard tuft of soil to release them.

In the end, we abandoned our direct path, and instead began to sweep a large arc, skirting around this seemingly-endless area until eventually emerging, and once more attempting our westwards trek.

Then we had a second puncture. Fortunately, we still had one more spare between us: but we knew that we couldn't now afford to lose another tyre. It was also late afternoon, and sunset wasn't all that far away.

Still, we continued to make good progress, if with a slightly-forced enthusiasm, until, just as it was getting dark, we approached a large, wooded area; to be precise, a stand of stunted acacia thorn, set so close together that it was extremely difficult to weave in and out of their trunks. There were points where we couldn't find a way ahead at all, and everyone but the drivers had to get out and walk ahead of the vehicles, lighting the way with dim, battery-powered torches, searching for enough space to get them through.

We'd just about come to the inevitable decision to stop and wait out the night when, through the trees, we saw the flickering glow of firelight. Heading moth-like towards it, we found a group of nomadic Fulani seated around their campfire, with their cattle grazing nearby. I don't know who was more surprised to see the other! Our drivers knew enough of the men's dialect to be able to explain where we were attempting to go, and the herdsmen gladly volunteered to show us the way out to the main road. That, surely, should have been the end of our problems: but the Fulani could not only run fast, but had no idea of the widths which the Land Rovers needed in order to follow. We lost sight of the tribesmen almost immediately, and had to wait a good while until they, more than a little puzzled, returned to find us again.

But in the end, we did reach the edge of the trees: and there ahead, once more, was the high-grass plain. What's more, right in front of us was Konan Kuka, with a huge, bright moon just rising above its pointed peak. It was a beautiful sight; and a navigational blessing. Keeping ourselves pointed towards the mountain, we at last found ourselves back on the road. From there, it was only a half hour's drive back to our base, which we finally reached around midnight.

Once again, nobody seemed particularly perturbed by our late return. Nobody, in fact, was still up and about at all. Dinner - complete, of course, with a most enjoyable Claret - had finished ages ago, everything had been cleaned up and put away, and everyone had long gone off to bed.

Hungry and exhausted, so did we.

Soon afterwards, CDC brought in a man from their Lagos office, engaged specifically to liaise with government officials on our behalf. Jonny Ngwube was a large and thoroughly charming fellow. He was also, it seemed to us, a somewhat odd choice, given that his Biafran tribe had, to simplify a terrible conflict considerably, only just a few years earlier been engaged in full-scale civil war with several others throughout the country, including those in whose territories we currently were. Fortunately, everyone now seemed to be on reasonable terms.

One of the first outcomes of Jonny's 'connections' was the arranging of a visit to a tomato puree factory. This, it was claimed, would allow us to see how well Nigeria handled general industrial production: essential if a successful sugar mill was to be built. Our group arrived to find a sizeable modern, functional-looking building set amidst a huge acreage of prime growing land: yet we could see only the sparsest scattering of actual live tomato plants, and no-one seemed to be working upon the plantation at all. We went inside, and immediately recognised the appropriate washing, grading, sorting, crushing, and sterilising equipment necessary for processing the fruit and concentrating its pulp: only none of that was working, either. Instead, standing wherever they could fit around and in-between the still-shiny machinery, we found a large group of men, each armed with a bucket. Into these, the workers firstly poured a quantity of dried tomato powder - imported powder, I soon noticed, from the large boxes' packaging - before adding water and stirring vigorously. The resulting orangy-red paste was then emptied out into the packaging machine - the only thing still in use - which in turn filled its small metal cans, fitted their lids, and sent them off to the stores.

As was common in many 'developing' countries of the time, the factory and initial land preparation, together with its first bumper crop of plump, juicy tomato plants, had all been provided and managed under a foreign aid package. But apparently - typically, I must sadly add - as ongoing management assistance and supervision had declined, been withdrawn, or quite possibly, as I subsequently encountered on more than one occasion, been outright dismissed, there simply weren't enough skills and knowledge left to allow continued, en masse cultivation.

Overall, it didn't leave us feeling terribly inspired.

Shortly after that particular trip, for the first but by no means last time, our stocks of diesel fuel, serving both the vehicles and generator, began to run dangerously low. No fresh supplies had been available in the local village for some time, no delivery was expected to the area in the foreseeable future, and we'd already heard that none of the towns to the south had any fuel left, either: all round, a not-particularly-unusual Nigerian occurrence in those days, for no overt, or certainly external or uncontrollable reason which we could ever understand. To make matters worse, our larder was also fast becoming severely depleted. (The imbibable stores had, I recall, been well planned for in advance: but, apparently, man cannot live by the vine alone.)

Our best chances seemed to lie north and west. There were more European settlers there, and things were likely to be better organised; or so we hoped. So we sent Jonny and Stuart off in one of the Land Rovers, its open back full of empty diesel drums, its tank holding sufficient fuel to reach their destination, but not enough for the trip back. Desperation or inventive incentivisation; I can't remember which.

Before setting confidently out, Jonny asked the group if there was anything special which we'd like bringing back. Jokingly, I suggested that some smoked salmon would be nice. We then bade the pair goodbye, trusting that they'd manage to return before our remaining fuel and food ran completely dry.

Thankfully, their hunt proved successful in very short order: they came back only two days later, the drums full of diesel, all other available space piled high with crates of assorted foods. The very next morning, therefore, in celebratory mood - and, contrary to whatever Mrs Mangmay might ever now think, in proper British style - we ordered up a full fried breakfast: eggs, bacon, sausage, and a slim square of frizzled brown stuff.

The cook was summoned, and we demanded to know what this latter item was.

It was something, he said, which he'd found in a package in the freezer.

Jonny absolutely did his nut! "That," he rose to his full height and bellowed, "Was prime smoked salmon. It should always be served cold."

"What? Not cooked?" The boy looked utterly mystified at these strange foreigners' customs.

Jonny, kind-hearted as ever, had intended us to have a pleasant surprise for dinner. Well, we did: just not quite when, nor in the way that he'd planned.

The Numan study ended, but Ned was quickly relocated over to the west of the country, to a spot midway between the small towns of Mokwa and Bida, a short (in African terms) distance north of the river Niger. My job, over three or more trips a year, was to supervise his work and budget.

We'd been summoned to examine another vast area, this time part of the Sunti Plain. This, like our previous dry swamp, used to be inundated every year by the great river's headwaters, but their flow was now regulated by the newly-constructed first Niger dam (the Kainji dam, as it's now called). The land was already looking ready for agricultural exploitation, and they wanted to know if it would grow cane.

Two bungalows had also been provided here, one previously housing the district road engineer, a post lost after the end of British colonial rule. In contrast to Numan, these brick-built, two-bedroomed properties provided not only very pleasant, but most luxuriantly-spacious accommodations for our much smaller 'team', Ned occupying one by himself - with me in the other bedroom when I came out - and giving the other to the 'houseboy'.

The management of T&L's existing Ilorin refinery, far the other side of the Niger, had also lent us one of their latest, top-of-the-line Range Rovers. It was, I must say, extremely comfortable for making the six- or seven-hour journey back and forth to Lagos. Despite all claims as to its specifications and intended abilities, however, although useful for reaching the edge of the Sunti plain, some forty or so miles away, it wasn't exactly the best vehicle for going off through the bush. But that was what we had to do, and it certainly wasn't the worst case of 'needs must' which I ever experienced.

We had to adopt pretty much the same approach when it came to planning our journeys. Ned and I had managed to construct a composite map of the area by piecing together some old aerial surveys. It soon occupied a whole wall in our bungalow's lounge, and was, to our knowledge, the only full mapping of the area ever then undertaken, its terrain and passable routes updated each and every time that we ventured off deeper into the plain. We soon discovered that, in addition to the huge swathes of land, there were also many lakes in the area, some of significant size. They had to be named, of course, if just to identify them on our maps, so we took as our reference point a suitably-modified English Lake District: and thus were formed Tatewater, Lylethwaite, Edwin's Tarn, Fendermere, and so on.

I sometimes wonder if those names still exist.

We also had to establish and plant up a trial farm, assessing which cane varieties would grow best. Permission had been gained from the Nigerian government to obtain whatever was required to equip it, so Ned and I soon found ourselves wandering blissfully around a large, US-aid-funded workshop at nearby Bida. Each of its several sheds contained many orderly rows of clean, reasonably-new-looking machinery: dozens of crawler and variously-wheeled tractors, harrows, rippers, ploughs, and all imaginable kinds of cane cultivation tools.

Talk about two children in a sweet shop! We'd just about decided what to take from this heavenly selection when the workshop's manager, a local man, wandered over and offhandedly mentioned, just in passing, that none of the available items actually worked. The first, for instance, he responded to our more specific enquiries, didn't have a generator; the second lacked its oil filter; the third had a fuel pump missing; the fourth needed some bearing or other... And so it went on.

Surprised, we asked why the missing parts couldn't simply be replaced with ones from other machines. Apparently, came back the shrugged reply, there probably wasn't any real reason why not. So back along the rows we went, searching for the requisite parts. Soon enough, we'd managed to assemble a complete and fully-functioning crawler. The same procedure then followed for a wheeled version, the only difference being that this time, still lacking an oil filter body, we later had to make do with an empty, discarded fruit can. (Once delivered, by the way, both tractors, together with everything else which we'd taken, continued to work perfectly well for the entire two years that we were on site; although the fruit tin did eventually have to be replaced.)

A major feature of my many visits to Sunti - as with my site visits in many other countries, of course - was the not-infrequent long journey down to Lagos and back. Such trips were, on the whole, nothing but boring and tiring, enlivened only by anticipating, and then counting, the latest accumulated collections of crashed trucks and buses which would inevitably be encountered at nearly every bridge en route.

These latter structures had all been built in a time when the roads were still only single track; track probably being a pretty operative word, too. Now, of course, the roads' widths had been increased to the standard two lanes, and they were all - for once - equipped with good tarred surfaces, enabling drivers to proceed along at a good speed, and with at least - as far as these things ever went in such places - some expectation of safety.

Except, of course, when negotiating the bridges.

Perhaps it was all down to some form of natural, innate arrogance or obstinacy - and many Nigerians certainly possessed their good share of both - but the local drivers seemed to love nothing more than a good game of 'chicken', each assuming that he held the right of way over any and all oncoming traffic, no matter what its position, type, size, or pace, and clearly holding no notion at all as to the generally-sensible concept of giving way. Hopefully, usually, one or both of the drivers would, at the very last possible moment, execute an emergency pull-off into the scrubby bush which ran alongside the rising road surface; and then, like as not, straight on down the steep embankment which it covered.

To us, it all just seemed vaguely amusing: until, one day, we came across the remains of a long articulated truck, atop whose load many of the locals had, as was the normal practice, in exchange for its driver's usual small monetary levy, also been riding. It, too, had driven off the road at one of the bridges' approaches, where it had then hit a group of large trees and - doubtless aided by some kind of flammable, and not so well protected or secured cargo - burst into flames. The heat had charred the trees right down to the ground. The humans had fared no better.

Another delight of my regular drive was crossing the bridge over the Niger. The wide river interrupted the main north-south route to and from the capital city, and the sheer, and still-increasing, volume of trucks, busses, and cars using it had by then far outstripped the old ferry's capacity to cope. There wasn't, however, yet an alternative road crossing in place: and so, most inventively, someone had come up with the bright idea that if the road could just be diverted a little, then the existing long, curving railway bridge could be pressed into service to carry other traffic as well. It hadn't, of course, been built for that: but with no further adaptation needed than the simple laying of an appropriately-solid, absolutely no unsecured logs and branches involved, deck in-between and outside of the tracks, flush with the top of the rails, the problem was quickly solved.

The road inevitably began to be used by even more vehicles. The congestion only increased again. But the rail line remained single-tracked, and the bridge's girders extended about two yards upwards on either side of the rails, enabling only one line of traffic to traverse the river at any time in any one direction. With admirable foresight, control and direction of the otherwise-resulting chaos had already been placed into the hands of two groups of policemen, each permanently stationed at either side of the river, both able to contact each other by telephone to determine which stream of vehicles could proceed in whichever direction they might jointly or severally choose. It wasn't at all uncommon for a delay of an hour or more to be endured if one was unfortunately caught on the opposing bank from where the main traffic was coming. It was also extremely common - nay, expected and effectively essential - for huge bribes, known locally as 'dashes', to be paid up-front to the police for the express purpose of obtaining priority. Of course, as these were equally likely to be paid on both sides of the bridge at the same time, there was invariably only one set of ultimate winners; financially, at least.

Matters became even more complicated when the police were notified that a train was coming. Since it could take a considerable time to clear all road traffic from the bridge, long 'train approach' waits became standard, and horrendous queues could very quickly accumulate by the time that the train had actually passed. In the meantime, many more dashes would have been paid to assist in the subsequent prioritisation of the vehicular flow.

Road travel was also very time-consuming during the last section from the airport into the city centre. It could often take up to an hour and a half just to cover ten miles; and there weren't even roadworks in sight. One time, with Ned and I having been driven down by one of our Nigerian drivers - by far from always the case - the stationary traffic jam could be seen stretching out so far in front that it continued way past the contractor's office which we needed to visit, itself still several hundreds of yards in front of our current position. So we simply stepped out of the Range Rover, told the driver to wait at the office once he finally got there, and began to walk. Our meeting had already been concluded, and we'd just returned down to the building's reception area, as our car turned onto the forecourt.

This last visit, incidentally, wasn't only necessitated by business matters, but also doubled as yet another of our many emergency 'hunting trips'. Supply shortages had again bitten, so Ned and I therefore loaded up the Range Rover with all of the various foods, equipment, and other necessary goods which we'd been able to buy in Lagos, and sent it back to Sunti without us. It hadn't been difficult, nor expensive, to hire a four-seater aircraft, so we arranged to be flown up to Mokwa, an old CDC plantation not far from Sunti, and with whose solitary remaining resident European member of staff we'd had occasional dealings.

Only once in the air and well on the way did our pilot happen to mention that he'd never actually been to Mokwa before, and, even with the aid of a rudimentary map, wasn't entirely sure where it was. We, fortunately, had a fair idea, and so, navigating via the roads below us, directed him straight there. By then, however, it was getting dark: the taxi battle out to the airport had been even worse than usual, and we'd ended up taking off quite a bit later in the afternoon than originally planned. We flew over the plantation, trying to identify the basic, unlit runway: but we could only make out what at best appeared to be an extremely short stretch, which the pilot - quite sensibly, in our opinions - firmly stated that he was not willing to risk. He was all for returning to Lagos, but we persuaded him to divert instead to another sugar estate, Bacita, on the south bank of the Niger, about two hours' drive from site. We already knew its general manager, another European, as they were selling us our seed cane, and it was indeed he who, as soon as we'd touched down, immediately drove out to investigate the unexpected arrivals, graciously gave us all a bed for the night, and the next day drove Ned and me back to our base.

We also found out why we hadn't been able to see the full Mokwa strip. It was actually, as it turns out, perfectly adequate in terms of overall length. It was just that whoever had first laid it out had, for some unfathomable reason, left it with a rather sharp bend right in its middle; and we, in the dusky light, simply hadn't noticed that it continued on around the corner. As, you know, runways normally do.

We were still overseeing the Sunti trials when the Nigerians decided to modify their road traffic laws. The authorities' first aim was to reduce the amount of vehicles going to and from the airport; or, at least, using the main road which led there. They therefore mandated that cars with even-numbered registration plates should use the road on even-numbered dates, and those with uneven numbers on uneven dates. It was a simple plan, and ought to have been extremely effective.

Except that they didn't reckon with their people's native ingenuity. To a man, and probably woman, the entire road-using population promptly went out and purchased a second set of number plates, whether genuine or not, and each morning merely swapped them over to suit the date.

Nothing much, consequently, changed at all.

The second alteration was rather more serious. Nigeria was surrounded by Francophone countries, all of which drove on the right. Nigeria, on the other hand, as a former British colony, had held fast to the good old code of keeping left. Someone must finally have thought it better to share commonality with the bordering states, so an edict was accordingly issued to such an effect. On the first day of the new system, it stipulated, all trucks would now drive on the right-hand side. With their drivers having by then become accustomed to the new ways, it further stated, all cars would change sides on the second day.

But - and it was a very big but - the new statutory instruction utterly failed to mention that no cars would be allowed onto the roads during that first, fateful day. Naturally, therefore, the trucks duly - mostly - obeyed the new rules; and, since no-one had informed them to the contrary, the car drivers continued to abide by the old. The collision statistics can only be imagined: and, of course, no driver could ever be said to be at fault, as each was still correctly obeying the highway code.

Things were apparently somewhat better by day three: except that many drivers then claimed never to have been informed about the changes at all, and, in any event, their reflexes instinctively, and unsurprising, still made them 'go left'. It took quite a long time, I understand, before things did eventually improve back to the everyday - utterly chaotic - norm, and its accompanying standard - exceptionally high - accident rate.

Towards the end of the Sunti trial farm project, I was granted the honour of meeting personally with the government's minister of industry. I wanted to discuss the possibility of T&L also putting up a processing factory there, and, at some point in our conversation, I suggested that, should this go ahead, we would be prepared to provide all of its initial staff. This wouldn't just be for the commissioning process, either, but for a period of up to ten years thereafter, during which time we'd additionally agree to train up suitable locals to take over its future running.

Memories of tomato paste plants notwithstanding, it seemed to me to be not only sound commercial practice, but a very good negotiating incentive. To my utter astonishment, however, I was immediately told in no uncertain - indeed, rather indignant - terms that this proposition was not at all acceptable. Furthermore, it was entirely unnecessary. The Nigerians, you see, being "very clever people", would require, at most, only two years to obtain the necessary skills to manage the entire project themselves. No further outside assistance of any kind would therefore be required.

Without casting any particular aspersions, this, in my by-then considerable experience - gained not just in Nigeria, by any means - didn't sound at all encouraging. In fact, as a direct result, although other sugar factories were subsequently built in the area by several other companies over the years to come, we soon decided that we weren't particularly keen to go any further with the proposal at all.

Events but a few months later served to reinforce our beliefs. Canadian water engineers had been contracted by the government to manage the new Niger dam, and had been controlling the river's flow so successfully that the previous floodplain, including the Sunti area, had dried and become ready for its planned development well ahead of schedule. Indeed, our own trial crop was already nicely into its second year of growth, and now standing at some eight to ten feet tall. Things, I thought, back in my UK office, were generally going well.

The telex from Ned was urgent, if extremely cryptic. "Fishing in cane fields," it read, "Now particularly good."

He didn't need to add any encouragement about packing my rods and coming out soonest. Something, clearly, had gone badly wrong. But, for reasons of commercial or political sensitivity, I presumed, he obviously hadn't dared sending an open message to any further explanatory effect. Only after several days of prolonged foreboding, therefore, anxiously awaiting the eventual arrival of a first-class airmailed letter, did I finally find out that, literally overnight, the entire Sunti plain had been flooded to such a depth that all of the cane was entirely underwater. The deluge had soon receded, but when Ned had gone out to inspect the damage, he'd found the cane fields entangled with numerous fishing lines. The ingenious locals were certainly never shy of an improvised approach when the opportunity arose.

It transpired that the Nigerian government had, in yet another demonstration of their great 'clever people' style of management, suddenly decided to renege on the dam contract. Packing the Canadians straight off back home, they'd immediately handed the project's day to day running over to their own people. The seasonal rains then began in the upper reaches of the Niger basin, and the dam's new operators, for reasons which we could never really fathom, barring either some gross misunderstanding of the dam's purpose, and certainly operation, or some strange kind of desire to return to the old, traditional, natural way of things, simply opened all of the spillway gates to their fullest extent. Straight away, the entire contents of one recently-created reservoir rushed off down the river, and out over all of the newly-reclaimed, newly-planted lands below; just as they always used to do.

I also rushed out to assess the damage. With the government's managers remaining firmly in command of the dam, and without their apparent willingness to accept, or was it just pay for, advice or assistance from either the Canadians, ourselves, CDC, or anyone else, it just wasn't worth us carrying on.

This wasn't the only downstream re-flooding that ever happened, by the way: and I must add that it wouldn't necessarily always have been wholly the fault of those newly-installed, and perhaps even presciently-over-cautious, managers. The problem of heavy rainfall always seems to have dogged Nigerian dam projects since that time until this day - and sometimes in other African, Asian, and American countries, too - the need to open spillways, or sometimes natural over-spillage itself, often causing significant or even devastating damage to land, property, and life.

We still couldn't escape the country without striking further troubles. Once again, they came in the all-too-familiar shape of food and fuel shortages. This time, however, the overall scarcity of things got so serious that we had to revert to living off the land. Even our lonely friend at the old Mokwa plantation couldn't help us when we turned to him for assistance: and he did by then, despite the counterbalancing loan of a slightly wonky airstrip on more than a few occasions, owe us quite a significant favour. After all, CDC had utterly failed to maintain their original staff facilities to any decent standard at all, so it had been left to us, on one of our various trips out from the homelands, to provide him with new set of pockets for their club house's full-sized billiard table; and also to transport out for him, free of otherwise hefty carriage and import charges, the latest in not-actually-very-lightweight-at-all - or so I was reliably informed - carbon-fibre-shafted golf clubs, in order that he could employ his remaining contracted time improving his drive on their nine-hole course.

What a difficult - and different - in-country life he'd had!

Still, even without CDC's fine provisions, we managed to sustain ourselves remarkably well, dining on such exotic local delicacies as freshly-caught fish, river terrapin soup, and, for a few days, roast rib of very large snake. (The latter, I should clarify, before you gain entirely the wrong vision of self-sufficient, trap-setting, prey-chasing agricultural engineers running amok with their whirling cane knives, was one day enjoying a rather sluggish start to its morning in the middle of a sun-warmed road. Our Range Rover, coming round the nearby bend, just didn't have time to brake.) It rather resembled small lamb cutlets, and, yes, tasted somewhat like chicken.

New supplies of diesel, at least, did finally reach us after a couple of weeks. It wasn't difficult to decide that it was time to pack up and use them on one final trip back to Lagos, where I bade goodbye to Ned for the time being, and to Nigeria for good.

Or so I fervently hoped.

9

AND OTHER AFRICAN ANTICS

This might have been the end for Nigeria. It certainly wasn't the end as regards West Africa, though; nor the rest of that great continent.

I'd visited Sierra Leone once before, in 1976, advising on the proposed creation of a large sugar estate near to its southern boundary. The area was totally unsuitable, so things had very quickly come to nothing.

I say nothing, but there must have been some residual good feeling created, as I received a postcard in December of that year from one of the ministry's planning officials with whom we'd worked. He firstly sent his apologies for not having been able to see me off when I left, but he had suddenly been "hit hard by this notorious disease of malaria". Then he thanked me profusely for the "lovely small parcel" which had apparently been sent to him on behalf of my team. (I don't now remember what or why, but doubtless a little kindness went a long way, and we must have seen a deserving reason.) He had, he informed me, immediately taken it to his "old mother in the village". He went on to say how much he had enjoyed "every moment of my stay with you people", and to tell me how much he had learned from us. Finally, he assured me that everyone was looking forward to hearing from us very soon, as they were "very keen to establish sugar estates in this country".

Thus was the way that things were sometimes done. Not a formal typed letter, not a telex, not official headed paper: just a hand-written colourful postcard, an annotated cartoon map of the country marked 'Dedicated to His Excellency The President' at the bottom. It was a rare and lovely gesture of appreciation by someone whose life was a lot more difficult than, and different to, my own; and you didn't overlook those when they came your way.

And keen they definitely were, for, a couple of years later, they asked us back again: this time to search for other, smaller-scale sites. Our remit was effectively wide enough to cover every single potentially-viable area of the country, and I was asked to lead the team, with Ed Alursin acting as our soil scientist, and Ned Fender our general Africa 'old hand'.

The outwards journey tried its utter best to compete with previous experiences, our plane developing some sudden mid-air "technical engine difficulty" which forced us to make a hurried, unplanned stop in The Gambia, at an airfield still no more than a wartime strip, made of perforated metal plates pressed into the grass. Once safely landed, the apologetic captain then informed us that, due to "peculiar radio signal conditions", he couldn't currently contact either London or Freetown, Sierra Leone's capital city. All passengers would therefore have to be bussed over to a local hotel to await a relief aircraft, once he could arrange one.

It actually flew in the next afternoon - not really too bad a delay, considering - and onwards we then went, arriving at our intended destination only to find that - quelle surprise! - no-one had been sent to the airport either to investigate why we'd never appeared as originally planned, nor to enquire about, and thus reschedule, and thence to await our collection in accordance with, our rearranged arrival. The airport itself was located quite a way to the north of Freetown, all the way across the other side of a very wide estuary-cum-bay, in fact, and rather inconveniently accessible only by a lengthy trip on a small, rather dilapidated-looking car ferry. But we managed, after a considerable wait, to summon up a taxi, and eventually made our way across and into the city proper, where, for want of direction to anywhere else, we checked in to its biggest and most expensive-looking hotel.

We were finally able to ring through to the Ministry of Agriculture the following morning. Our local agricultural counterpart, Saul, arrived shortly thereafter, promptly driving us straight over to his "head office", a long, wooden, ramshackle-looking building set atop a small hill. It might well have looked almost stately in the old days; now, it just sat there, sadly decrepit. We examined what available maps they held of the area - not a great deal in either number or coverage, and lacking in far too much detail - decided upon a rough work programme, and then went to find the car which the government had agreed to provide.

It was, we were pleased to see, a large old American model, with plenty of room for us all - including the unexpected addition of a driver - plus provisions, luggage, and equipment. Its tyres, however, were far from a pleasant sight. They weren't just worn, but well the other side of bald, with most of the canvas showing through: definitely not something upon which we'd ever contemplate setting out upon a rough, countrywide journey. Yet when we politely pointed out the need for some replacements, we were promptly, if equally politely, told that the government simply hadn't the funds to do so. They had, however, already arranged for us to be advanced a cash sum to fund our forthcoming travels; and so, heading off next to the Central Bank, we successfully - after the usual lengthy delays - collected the promised money, and straightaway spent nearly all of it on a new set of tyres.

This unplanned initial outlay then led on to another problem, since our group, when not in the capital, was to be accommodated in various pre-arranged self-catering premises, and we were now left with very little to cover all of the required food and other necessary items which we needed to take with us, not only for ourselves, but for Saul and now the driver too. Luckily for us, Ned already possessed extensive experience of living in a good many remote places with extremely limited resources - wild snake and terrapin notwithstanding - so he quickly devised some impromptu advance menus, and calculated the minimum quantity and quality of edible and other supplies which we'd have to acquire.

And so, at length, we set off. Quite surprisingly, our several weeks of travels - indeed covering practically the whole of the country, excepting those parts where I'd previously been - passed wholly without incident. Thanks to Ned, we managed quite adequately within our extremely limited means, and, in-between our more far-flung forays, we even managed to spend a couple of enjoyable weekends in Freetown itself, lazily passing the time on Lumley beach, and eating out at a cheap Lebanese restaurant which served excellent curries for Sunday lunch.

The final task of that particular trip was to meet with the country's vice-president. He was, I felt almost immediately, a man whom I couldn't altogether trust. Perhaps this instinct wasn't wholly without foundation, for, some years later, I heard that he'd been imprisoned for massive fraud.

But then again, corruption and financial mismanagement, in all of their forms, were simply an endemic part of the West African scene, from the lowliest porter or servant to the highest-ranking official. In truth, though, it was a feature so embedded throughout the vast majority of my working world, by far not confined to Africa, that I'd soon became quite used to it, the only real difference between the various countries being the time at which the bribe was actually expected - before or after the event; perhaps, even, in stage payments - how much would be required, if not just outright demanded, in what form - cash, kind, or contractual mark-up - to whom or across how many it would be allocated, and whether it was ever acceptably possible to circumvent the local tradition by adopting our usual do-it-yourself approach.

Somewhat later, another T&L survey was sent out to the country to look for yet more alternative cane growing sites. Medium-sized ones, this time; just in case we'd somehow missed them. Still, you can't possibly say that they weren't eager to continue exploring all their options. We'd included several of our younger staff in this particular study, enabling them to gain further field experience under the care of one of our senior agronomists. I was also asked to pay a visit, just to check how things were going.

The five consultants were billeted in a rather rudimentary bungalow, in a village near to a stretch of uncultivated land which was several miles wide. Soon enough, along with two local guides, we were all setting out in our Land Rovers towards its far boundary, from where, with the intention of walking the return leg, we sent the vehicles and drivers back.

After a few easy enough hours on foot, we unexpectedly came across what I can only describe as a serious gorge, its steep sides dropping about thirty feet down to a fast-running river, which also appeared to be very deep. Fortunately, straight ahead of us, we could see a bridge.

I use the term 'bridge' only in its most base and loosest definition, however, for this perilous contraption was constructed mainly from the stems of climbing plants, known in some parts of the world as Lianas, woven together to form rudimentary ropes. There weren't even any timber frames to support its ends, the suspension and walkway 'ropes' simply being tied around nearby trees on each bank. The footway itself, some twenty five or more yards in length, was just wide enough for one shoe or boot to be put down at any one time, and consisted only of a most insecure, and unsecured, series of thin, longitudinally-placed slivers of wood - half-branches, to be exact - easily able to slip sideways if said foot wasn't placed very carefully indeed. It also sloped severely downwards towards its centre, any crossing therefore necessitating an exceptionally-cautious descent to its swaying middle, followed by an extremely-precipitous climb out to the far end. To make matters all the more amusing - or not, as was most definitely my current view - gaining initial access first entailed scrambling up an incredibly rickety ladder, made from nothing more than thin tree branches tied together by creepers and vines.

The bungalow here was also, albeit in a different sense, considered particularly insecure, so we'd long deemed it prudent to carry all valuable documents, including passports, airline tickets, and traveller's cheques, upon our persons when going out into the field. Everyone had their own preferred means, methods, and locations: mine, like many others', were safely contained within a small zippered bag, slung casually over one shoulder by a long strap, whilst my other shoulder similarly supported the hard, shaped case of my far-from-lightweight camera. One of our party even favoured transporting his vital paperwork - along with a few local maps, it should be said - in a full-sized, leather-sided, hand-held briefcase: and thus he'd equipped himself on this day's outing, too. Lest it appeared as if we weren't taking our duties fully seriously, we were also laden with a couple of sturdy spades and a long, and very heavy, auger, with which to take soil samples from time to time; some, in fact, were already being carried back with us in large, hand-held plastic bags.

The bridge was, our combined engineering expertise quickly surmised, only strong enough for one person at a time: which in turn meant that no assistance could be rendered by a second, should it be required. Still, it didn't present any problems to our guides, who swiftly picked up the tools and bags, ascended the ladder, balanced the lot upon their heads, walked down the slope of the bridge, mounted the far side, and then pretty much jumped back to the ground; all without even touching the ropes on either side. Then, just to demonstrate how safe and easy it was, they both trotted back across to where we all, now rather unhappily, stood.

Then it was my turn. Well, I was the most senior man present, someone had to take the plunge - very metaphorically, I prayed - and I might as well get it over with. Besides, after Cirebon, I was surely an old dab hand... So up the ladder I went, document bag hanging down on one side, camera on the other, both hands running as lightly as I could allow along the guiding ropes, at first at shoulder height, then sagging more than proportionately as I progressed.

All things considered, it didn't seem to be too bad an experience after all. Not, at least, until I reached mid span, when the entire structure beneath me suddenly swung out sideways. Instinctively, I grabbed for the now rather low 'handrails'; and, as I did so, I felt the bag's strap sliding slowly but surely off my shoulder, down along my bare arm, and off over my outstretched hand. There was nothing that I could do to stop it, clinging on to both sides for dear life as I then was. Momentarily, I contemplated the mental picture, everything which I required to get me out of this country and returned back to my own now falling in slow motion down into the swirling white waters, before being rapidly swept away, forever lost.

And then I dared to look down. By some random stroke of luck, the little bag had managed to lodge itself between two of the slim slats which ran underfoot. But between my hands and the narrow walkway, there was no support at all; nothing, except pure thin air. Nor could anyone come to my aid. I don't recall anyone even shouting advice or making a suggestion. Summoning up all of my resolve and courage - not an awful lot at that point, I have to admit - whilst still tightly clutching the side rope with one hand - and after ensuring, with wonderful hindsight, that the camera's strap was fully secured both over my head and around my neck and body - I bent very slowly forwards, balancing precariously on one foot, desperately hoping that the whole ridiculous contraption wouldn't again start to slip out from under me; or worse. Gradually, inch by wobbling inch, I stretched my free hand forwards, reaching out until my trembling fingers could finally hook themselves into the bag's thin strap, currently hanging down over the edge. And then, very, very carefully, mindful not to let the long, dangling loop snag on anything else, I eased myself back upright until, with considerable relief, I could also pass that strap over my head.

The stiff haul up the slope to the far side seemed but a mere stroll in the park after that; the descent of the equally-decrepit ladder to ground-level safety an utter breeze.

My plight, of course, hadn't gone unnoticed by my colleagues still on the other side. But there weren't - disappointingly, I felt - any rousing cheers or congratulatory rounds of clapping forthcoming: just a silent huddle, hurriedly ensuring that their own variously-slung, strapped, clipped, and pocketed documents were much more firmly stowed. All, that was, except for he of the bulky briefcase, who was looking severely worried. But one of the guides soon solved his problem, proceeding, once again, simply and literally to use his head. Balancing the large, flat-sided case just as before, somehow managing to keep both hands free - and yet also wholly otherwise unused - he almost, to my undying shame, ran straight across.

****

Cote d'Ivoire - the Ivory Coast - was a vastly different environment. The customs and culture were French, the food was French, and, of course, so was the language. Proper grammatical French, too, as once demanded of all of France's colonial citizens, in contrast to the 'pigeon' English spoken in most formerly-British lands. In another contrast with our own past 'Empire', Frenchmen were still to be found amongst all classes of society, occupying any and all types of employment and position, the lowliest to the most senior, side by side with native Africans.

My initial 1973 visit, along with two other agriculturalists, was to look at a possible new estate site near to an existing one - actually the first, and by then still only in the country - at Ferkessedougou, up in the north. One of my colleagues happened to have a personal friend, an American, already working at 'Ferke', so we were very well looked after, and his local knowledge and contacts enabled us, even in a very small town, to dine out surprisingly well on superbly-prepared French cuisine, and to drink from a wide selection of high-quality French wines.

CDC would surely have loved it.

It wasn't long before we'd identified a suitable area of land. But, at the instigation of the country's president - this, after all, being a government-sponsored project - we also investigated an alternative site. Given its rocky and generally less desirable soils, there was, in our combined opinions, much less to recommend it. Yet, lo and behold, when it came to the subsequent feasibility study, it was only to the second area that we were directed. The president's mother and her family apparently lived in the locality, and their village had been assured a vibrant future as the new development's economic hub.

During this later study, and then the project's subsequent development and equipment-procurement stages, I often stayed at an hotel in the capital, Abidjan. For the times, in contrast to many other African cities, it was a very modern and pleasant place to stay, with tall skyscrapers, broad, well-paved roads, and many high quality, European-standard shops. The restaurants, needless to say, were excellent.

Late one night, however, on returning from a brief trip 'up country', I found huge, chanting crowds filling the streets. It wasn't, to my relief, some sudden revolutionary insurrection, nor a strike or protest march: merely a very important international football match being played in the vast stadium just around the corner from my hotel. I managed, at the start, if with considerable difficulty, to push my way through. But then I found myself confronting a solid cordon of stout and serious-looking policemen, who stubbornly, despite my rather out-of-place attire and appearance, refused to let me pass. It took a lot of persuasion to assure them that I was only a visiting businessman trying to get back to his room, and held no interest whatsoever in attempting to gatecrash the game. Derby County weren't playing, after all, so I really couldn't see the point.

Still, at least in that part of Africa, they had their football crowds well under control.

Another morning, though, things didn't go so well. Two of us had set out very early to meet another T&L man who was flying in. Half way to the airport, at the outskirts of the city, our taxi was stopped at a police barrier which stretched the width of the road. They demanded to see our 'passes', our permissions to be out and about at this hour.

We didn't have any. We didn't even know about the need for any. There was, it seemed, some kind of night curfew in operation, and although it was by now full daylight, the restrictions still, by law, had half an hour to run. We didn't know that, either, and the taxi driver certainly hadn't seemed at all bothered about it, nor ever thought to mention it to us.

The police then instead demanded to see our passports: but, given that we weren't planning on travelling anywhere, we'd left them safely stored back at our hotel. Wondering what else to do, I pulled out my wallet and produced my old paper UK driving licence, onto which, in case of need in foreign countries, I always clipped a spare passport photo. The policeman took it from me, unfolded it carefully, opened it out fully, looked at it from both sides, first one way up and then the other. He obviously couldn't read a word of it, and clearly hadn't ever seen such a document before. I assured him that, in terms of formal identification, it was an official British government pass. I might even possibly have intimated that, being such a large and lengthy piece of paper, with lots of small writing, pictures, and many to-be-stamped-or-written-into-only-by-duly-authorised-officials spaces upon it, it was quite a high-level, rather important one.

He didn't argue further, and quickly let us through.

Though I made many more visits to Cote d'Ivoire over the years, I never once encountered this kind of problem again. But I always made sure to carry my passport on me. And my driving licence. Just in case.

****

By name alone, the Gold Coast, as it was formerly known during British rule, must surely rank alongside, if not surpass, its near neighbour.

The contrast between life and general conditions there and in the Ivory Coast was certainly obvious as soon as I landed.

Ghana was, frankly, dreadful.

My first 1970s visit, along with another T&L man, James Winters, wasn't too bad. The Ambassador Hotel in Accra was reasonably comfortable, and served passable meals. But even by then, not too long after the overthrow of the Nkrumah regime, the overall state of the roads, together with many everyday general services, was screaming out for attention. In its colonial days, Ghana had indeed been one of, if not the richest state in the whole of the African continent: but its first post-independence administration had bled it dry, and it was now fast heading to become one of the poorest.

The two existing sugar estates and factories which we assessed weren't faring any better, looking very run down, obviously ill-maintained, and with production at a virtual standstill. The irrigation pumps, upon which cane cultivation depended, had long ago ceased to function, and the overall mechanical state of the mills was nothing short of appalling. Even the renowned Volta (or Akosombo) river dam, although a fairly recently-completed project, was in trouble. It had been constructed to provide water and power not only for Ghana, but for many surrounding areas; sufficient electricity "for Africa", as was at some point claimed. We were given the opportunity to tour it, but already, instead of being assured by a beacon of hope and modern progress, we once more found a terribly sad and sorry sight. Few of its water pumps still worked, the turbines couldn't turn due to a lack of parts and repairs, and the sluices just poured water straight down the river without powering any generators at all.

We also had the chance to look around two large, round forts which we'd passed on the way to one of the coastal mills. They'd once been used to hold captured slaves, prior to their shipping off to the South and North Americas and West Indies. Yet another, if for very different reasons, unpleasant experience for our European eyes.

A few years later, T&L were invited to tender for the restoration and management of a co-operative general farming scheme in an area known as Tono-Vea. It was right up at the far northern border with Upper Volta, soon to become Burkina Faso: words which, apparently, mean 'land of incorruptible people', though I never had the opportunity to put such a statement to the test.

Ghana itself had, by this time, undergone another change of government, brought about by yet another uprising-cum-coup.

The fledging country, like so many others in Africa and in similar, post-colonial positions, did seem to be quite keen on them in those days. This one was led by a certain Flight Lieutenant Rawlings. The country was in utter chaos.

Nevertheless, we went and talked with the various new governmental departments responsible for the area, met with the new people who were running this particular project, and visited the old, now-abandoned lands which had previously been developed with the help of Yugoslavian aid. It was all, once again, a horrible mess. The irrigation systems had long since dried up, and the two small dams which should have fed water to the farms were wholly inoperative.

But we went ahead and secured the contract anyway, soon recruiting six site managers. And then, before we could even get them into action, there was yet another change of government - less of an outright coup, I understood, and more of an agreed handover - this time based around a communist-style administration. The new government did eventually decide to carry on with the project, although they imposed several strict restrictions on how we could deal with the farmers and labourers.

I paid several more visits to Accra whilst we undertook our work, always staying at the Ambassador, now the self-proclaimed "city's best hotel". I hated to think what the worst was like, as the Ambassador had, along with everything else, including the country's entire infrastructure, deteriorated very badly indeed since my first stay. There was no running water for either washing or drinking, food was extremely scarce, and of incredibly poor quality, and the power supply was spasmodic at best. (This latter problem wasn't at all helped by the common practice of people digging up whole lengths of underground electrical cable in order to sell its copper on the black market: assuming, of course, that they actually survived the also all-too common dire, usually deadly results of their blithely-placed chops and cuts.) Accordingly, the hotel's staff provided me with two bucket-fulls of water on each morning of my stay, to be used for the combined purposes of my general ablutions and the flushing of the toilet. I could only then obtain more by bribing the room boy: not, of course, a wholly-atypical way to go about things, in Ghana or elsewhere. Still, it was clear that the country's entire population - well, the vast 'ordinary' majority - weren't faring any better, so I supposed that I couldn't complain.

Despite the many obvious, and not-so-obvious, challenges of that time, I continued to work out of our new Accra office, recruiting and organising the site staff, and obtaining vehicles and housing for them. And then, before travelling up to site myself, I went along to introduce our new general manager to the minister of agriculture. The interview had been arranged many days in advance, but we were still forced to wait for a very, very long time upon an open, and consequently stiflingly-hot, veranda. Meanwhile, several various minor officials came and went, in and out; and always completely ignored us. To my great surprise, therefore, and against all mounting presumptions and preconceptions, when we were eventually called in to see the minister, he did nothing but apologise profusely for his staff's actions, continually explaining how he'd had absolutely no idea that we were even outside. He did, for once, in complete contrast to his staff and many of his fellow ministers, seem to be a genuinely-nice old bloke.

I wonder if he survived the next coup.

We'd purchased a brand new Range Rover for this project. It was a special diesel conversion - the first of its kind, as far as I could gather - given that, amongst its many other benefits, this was the only vehicle fuel likely to be found with any form of security in these types of conditions and countries; and even then, as Nigeria so often proved, security in no way equalled certainty. Our choice was definitely a wise one, though: during my second trip to Tono-Vea, the whole of the country's petrol-engined cars and trucks suddenly ground to a complete halt. Someone in the Ministry of Industry, it was later reported, no doubt upon receipt of some hefty kick-back, had purchased a large cargo of petrol, previously-condemned, and still heavily contaminated with sea water.

Despite our transport's relative luxury, however, the journey to site remained awful. The roads, previously laid with very good tarred surfaces, had now crumbled into near non-existence, and were often nothing but rutted mud tracks. They were perhaps a little better in the north of the country, but heading south, down to coastal Accra via the city of Kumasi (in colonial times the county's administrative centre), there was simply a greater volume of traffic. But it wasn't anything which we weren't already well used to, nothing stopped us being able to get on with our work at site, and I continued to visit every few months.

Amongst my many old papers, I recently came across a copied part of a hand-written letter which one of my colleagues from that project had written to another; his first progress report after arriving out, it seems, and I'd kept it for some reason or other.

He wasn't happy, to say the least. There was no control of anything, the local cattle included; very little had yet been had taken over into any kind of effective management; the much-promised demonstration farm had yet to materialise; there was considerable water wastage on site, and it was just passing straight down the two main irrigation canals, into an ever-growing hole which was now cutting a large gully in the bedrock, but the local water company had no diesel and no cement, so could do little to remedy the situation; they'd found a huge crack all along one of the dam's walls which they'd been inspecting, and it, too, needed urgent remedial attention. To do all of their land and site inspections, he and his colleague had been given the use of two old vehicles - if and once his driver ever bothered to return with his, current time and date unknown - but also only two gallons of petrol per day between them; and one car's windows didn't work, either. But he had managed to meet with the local MP and other local political figures, over a beer and a power failure in the 'club'. None of them seemed to have much purpose except to "sit, look and agree", but the MP did seem to be a "jovial enough character", and had offered to act as a conduit to parliament itself, which the team were now preparing to "exploit".

General communications throughout the project were poor, misunderstandings were rife - not least with his driver, apparently - people were all working in different buildings, and whilst they had been offered a "table" to work on at site, they had no diesel for power, air-conditioning was banned in all offices, and there weren't any overhead or pedestal fans provided despite the unbearably-high humidity. There was "inadequate" locking of gates on site, and those that were secured were able to be opened all too easily: consequently, there was some considerable "wastage" occurring, not to mention overt "strong-arm tactics" by local farmers. Stronger chains and padlocks were requested. Oh, and the weaver birds were eating all the rice: please also send bird scarers. And had he mentioned that there was no diesel on site? Because the local company were therefore thinking of returning the generators which ought to power the offices and site housing, including their water pumps, cooling systems, fridges and freezers. And the "cheap and nasty" site radio system wasn't up to the job, either: not only temperamental in the extreme, but the high frequency link didn't work, and had blown an intervening modem, and then its replacement, too, and servicing costs were prohibitively expensive, with spares even more so. Oh, and did Frank tell you that "our 'personnel' department has once again excelled in its efficient continued non-handling of overseas personal problems?" His "personal effects" had been sent out separately to the essential food which had been ordered for his stay, so surely could have been planned to have arrived at least a week earlier? As it was, they'd arrived in Accra the day that he'd left there, a week after he'd arrived in Ghana itself, and still hadn't yet been sent up on the lorry, because that was full of the water company's stuff instead. And as for food, the latest that he'd heard, via Frank... (There, my three copied pages ended). Still, "I'm only here for three months, thank goodness".

I never said that it was all roses.

My own normal routine was to travel to Ghana via Ouagadougou, in Upper Volta, from where I was collected and brought, quite legally, over the border to site. One day, though - it was, although I didn't then know it, to be my last visit - I landed at 'Ouaga' to find no-one waiting to meet me. I supposed that the normally-reliable driver had just been delayed for some reason, and so, after waiting around for a while, I took a taxi to a nearby hotel, had a meal, and put my feet up on the veranda. Late in the afternoon, correctly guessing where I was, the driver finally drew up outside, apologetically informing me that there hadn't been a traffic accident, nor a bridge or road collapse, flood, fire, mudslide, temporary blockage, log-jam, house-move, nor any of the other multifarious problems with which I was by then all too familiar: just another coup, of sorts, and only in northern Ghana. In any event, the local 'communists' had closed the border. My enterprising driver had therefore turned around and returned to site: but only for the purposes of obtaining some money, so that he could then bribe a border guard to let him bypass the official crossing via a small back road. We could, he proceeded to explain, if we hurried, take the same route back; assuming that the same guard was still on duty.

Fortunately, he was. Annoyingly for me, however, this unorthodox solution meant that I wouldn't have an entry stamp in my passport, an omission which I didn't particularly care to have to explain upon my eventual exit; nor, as now seemed quite likely, if stopped anywhere else and required to justify my presence. But with my driver paying but a brief return visit to the Ghanaian side of the border post, and another payment being handed to the appropriate man who held control over the stamping devices, all was once more, as far as things generally went there, well.

Some weeks later, a colleague and I were invited to have drinks with the local 'king', as the tribal chief styled himself. We were entertained within a large house, its many rooms oddly furnished entirely - and only - with sofas and chairs, whilst, amidst its gardens, sat the standard small, palm-thatched hut within which he and his family habitually lived. It was a strange set-up, and I never fully understood its significance and meaning. Still, he was a generous and friendly host, and during our conversation I happened to mention my recent border-crossing difficulties.

"Oh," he'd laughed, "If only your site manager had mentioned it to me. I would have come out myself to collect you, without any cost."

It transpired that he regularly travelled to Ouagadougou, and with his people and their native homelands spread out over both sides of this artificial border, he never bothered or needed to go through any official check post of any kind.

I nearly didn't make my final return to Accra, either. Travelling once more down what remained of the Ghanaian roads, miles from anywhere, a mighty clattering sound began to emanate from the front of the engine. The Range Rover's fan casing, it was clear to see, had decided to shear its bolts. But fear not, for I still always carried my trusty travelling stock of string, aided now by the odd collected bits of wire from here and there. We soon managed to secure the offending cover, preventing further fouling of the fan blades until we reached Kumasi, and located some proper fasteners of a reasonably-suitable size.

Not long after my return to the UK, there was yet another uprising in the north. This time, the new local politicians decided to invade our staff housing, holding everyone there under house arrest. No-one ever appeared to be in any real danger, but we couldn't sustain this kind of uncertainty, nor risk any form of further interference with the working or personal lives of our employees and their families, so we withdrew them all back home forthwith.

Our contract was nearing its end in any case, and, at my insistence, we'd already been paid in advance; and in full. So we, at least, had nothing to lose.

****

The early Portuguese sailors sent forth to explore the Gulf of Guinea seemed to be a little more inventive than some of their compatriots; if not necessarily so complimentary. They named their discovery after the word for shrimp.

Cameroun (or, in English, Cameroon) was therefore another country with a strong initial Portuguese influence; until it then came under German rule; until the end of the First World War, when the colony was subdivided between Britain and France; until it was then reunited, and finally granted its independence. Out of all of these confusing influences, the French seemed still to predominate whilst I was there, especially in Douala, the main point of entry, and its capital, Yaounde.

My first, mid 1970s visit was with a joint Belgian team, and was distinguished by nothing more than our final journey back to the capital before flying home. We were running behind schedule, so our Belgian comrades told the car driver to hurry. Having already demonstrated a certain French-style flair for fast driving, he needed little further encouragement to exceed all of his previous efforts. At least the roads here were of reasonable quality: if comprised entirely of compressed laterite. Clouds of brick-red dust were churned up in all directions as we flew along, not only coating all the car's windows, but soon permeating its interior as well. By the time that we arrived in Yaounde, we looked exactly like the pictures of 'Red Indians' which I'd seen in my childhood books; that, or several cases of very bad British 'pale-skin' sunburn. It wasn't just our faces, either: it got absolutely everywhere.

We arrived at the city's outskirts with just an hour to catch our plane, but in such a dishevelled and discoloured state that we severely doubted they'd let us on. A hotel room was very rapidly secured, and I still vividly remember standing under its shower, watching the red-stained water streaming down my body and circling off into the plug hole. We did make the airport, if just in time and somewhat damp: but our light-coloured tropical clothes, hurriedly stuffed into our baggage, took an awful lot of repeated washing once we'd arrived home, and always, to the knowledgeable eye, retained a slight pinky-orange tint.

I was back again a year later, now based in the northern Benoue area, near the river-port town of Garoua. This time, we were working as part of a large group put together jointly with our Canadian-based company: a very useful alliance, not least because, being from Montreal, they spoke fluent French.

The area of land under investigation was immense, dominated by a river valley which flooded every year. Crops, as usual, were grown in the fertile, and re-fertilised, ground which remained once the waters had gone. As was the case in many African countries around that time, a new dam was soon to be built to control this annual deluge, also providing year-round irrigation to the valley and beyond. Work hadn't yet officially begun, but the Chinese, constructing it on behalf of the Cameroun government, were already known to be on site. So we approached the relevant ministers, asking for further details.

And they, followed in short order by every official from the country's National Investment Society - the organisation ultimately overseeing the whole project - simply replied that they had none. The Chinese, all explanations then ran, had refused to divulge anything to them. This was, to be fair, fairly typical of Chinese aid projects in our experience: you just got whatever they provided, no questions asked, nor answers given.

None the wiser, therefore, three of us - Jim Villiers, myself, and another T&L engineer - spent the next several days touring the surrounding countryside and studying the available mapping. Then we went to visit the Chinese dam engineers' camp. We asked them if they could give us any idea of their plans, in particular the intended irrigation flows.

But they, too, replied in the negative. The relevant calculations had all been returned to China, they said. They only held the actual construction details; which they also, for some obscure reason, couldn't at that moment provide.

Still, we continued to talk some more, polite generalities mostly, and then we had some tea, and then Jim - always most adept at summing up a situation and arriving at a practical, civil-engineering solution - casually held up some off-the-cuff sketches which he'd just made whilst we'd been sitting there. These showed not only where and how the dam would be built, but provided estimations of river flows, water retention and release capacities, and several different irrigation options.

The effect on the Chinese was dramatic. Within minutes, they were scrambling around their various offices, and were soon handing round full sets of their own plans, calculations, and many other details, all of which coincided exactly with what Jim had just produced.

Thereafter, we all became great friends, and they soon invited us, along with the entirety of our site team, to visit and play table tennis with them as often as we wished: which, indeed, we later did. A sort of entente `ping-pong-tiddle-i-po`, I suppose: which, to non-Goons fans not of a particular age group, probably won't make one jot of sense.

****

Zaire, previously and once again the Democratic Republic of the Congo, is a massive country. Extending eastwards from a tiny, sea-bordered strip, it reaches more than half way across the continent, effectively encompassing Africa's entire middle section; its very south eastern corner, rather like the heel of the boot of Italy, forms an additional long, oddly-projecting 'pedicle', almost dividing neighbouring Zambia clean into two. The country's rivers and their many tributaries were also on a similar scale: long, wide, and often deep enough for large steamers to travel for hundreds of miles.

I went out along with Brian McCloud, an agronomist, to report on a huge expanse of land lying between the Kasai and Kwilu rivers, both of which flow into the mighty Congo itself. It was well over a decade since the country had gained independence from Belgian rule: but its general politics and governance were still, not entirely unexpectedly, far from in the best of health. In fact, given the very obvious numbers of Belgian residents busy getting out as fast as they conveniently could, things seemed to be fast declining. It soon became apparent that it wasn't only their former masters who were currently being suspected as very unwelcome supporters of the old regime, either: it was all 'whites', from wherever and whyever they came.

That last small issue notwithstanding, the government had still invited us to come. And so, on our first day in Kinshasa, the far-from centrally-situated coastal capital, our local agent took us over to meet an official of whichever department had thought that we might just have our uses. There, we were immediately provided with our very own military escort: a short, stout, army sergeant who spoke little English, but with whom, using a little basic French, we managed to make ourselves understood. He, we were informed, would remain with us for the duration of our stay, thus ensuring our safety; if not, in all likelihood, also reporting back on our activities.

We were also told that, in order to undertake our required flight on to Bandundu, some five hundred miles distant, we first had to obtain a 'Certificate of Permission' - some kind of additional internal passport, apparently - from an appropriate part of the airport through which we'd only so recently arrived. So Brian and I duly trooped back to the specified office, where we then found that, in order to get the required document, we each had to present a 'Certificate of Health', our internationally-accepted vaccination and other certifications not being regarded as at all sufficient for these particular purposes; or rather, we suspected, not occasioning the necessary payment which yet another piece of paperwork would require.

An appointment was therefore made for us to visit the airport's medical facilities that same afternoon. In the meantime, having the rest of the morning to spare, we decided to stock up on food to take out to site, not otherwise knowing what, if any, we'd later be able to buy; never mind, of course, if further certificates of hunger, or permissions to purchase, or some other similar documentations, might also be required.

We returned a few hours later, now carrying several bags laden with assorted canned and packaged goodies, and sat down in front of the doctor. He spoke surprisingly good English, and thus was quite easily able to explain that he was not at all permitted to sign such a medical clearance under any circumstances until we could prove that we'd been in the country for at least ten days.

Somewhat astonished, we, by return, pointed out that ten days was the total time allocated, by his own government, for completion of our survey.

We talked and explained and pleaded for a good while longer, but had clearly reached an impasse: until, that is, we happened to mention that we'd even just been shopping in the clear expectation of going onwards the very next day. The doctor, whilst conversationally enquiring what we'd bought, then reached into one of the bags sagging atop the table. He took out two tins of sardines, examined them briefly - professionally checking up on the nutritional qualities of our intended diet, obviously - and set them down on the table before him.

"These look very nice," he observed. "You know, perhaps, for such important men, working to further our country's prosperity, we can stretch a few rules. I'll have the certificates arranged straight away."

Thereupon, he took out his pad of forms, signed one for each of us, and wished us a good journey.

It was the cheapest form of corruption that I've ever encountered, before or since.

We flew off the next morning, along with our singular protective detail. Clearing local customs at Bandundu wasn't difficult, a Jeep and driver awaited us, and before long we were deposited in town at a scrubby sort of guesthouse. It was by now late afternoon, and our security escort seemed quite anxious to get off to his own quarters, leaving us alone to explore ours.

The rooms didn't seem particularly clean, but we'd seen worse, so we started to unpack, wash and change. We'd both, in fact, just finished when the sounds of a muffled commotion came from outside, followed by some louder shouting, then a heavy pounding upon our rooms' doors.

"Pack your things quickly!" a tall, white man commanded, when we poked our respective heads outside. "And come with me!"

The man, dressed in the typical European tropical style of the time, had a decidedly-foreign accent, but carried about him such an air of urgency that we obeyed without further question, fearing some kind of imminent full-blown riot, armed insurrection or uprising. Oh lovely, another coup!

Once outside, in the perfectly calm and normal-looking street, the two of us, along with our baggage, were swiftly bundled into his awaiting open-backed vehicle, and off we sped in a cloud of dust. As we went along, our unknown impromptu rescuer - the word kidnapper had also begun to cross my mind - finally explained, in the gravest of tones, that the place in which we'd been billeted was, in fact, the local bordello. But most fortunately, in his words, one of his staff had seen us arrive, made enquiries as to who we were, and reported their findings back to him: at which point, he added, he'd thought it his unquestionable duty to come to our aid.

We thanked him effusively, obviously, before very politely enquiring as to exactly who he was.

The abbot of the local German monastery was the basic gist of his reply; followed by an invitation, pretty much an insistence, that we henceforth stayed with him and his brothers for whatever length of time we wished.

Which we indeed then did, spending a very enjoyable time there, whilst being allowed to contribute nothing more to them than our few bags of food.

Safely re-accommodated, we passed the next two days driving across the designated study land, stopping now and again to look more closely at the soils and vegetation. Come the third day, and we'd reached our fairly standard decision that, in order to gain a better appreciation of conditions, we needed to walk a few miles across the grassy plain; grass, in this sense, meaning the usual high, dry, tropical type. So we packed up some food and water, parked up the Jeep, and started off in the direction which we'd plotted beforehand. Our plump little bodyguard had by now managed to track us down again, and thus felt compelled to come along, since, in his mind, we were doubtless going to abscond yet again, or otherwise do something terribly seditious.

Brian and I, neither of us particularly tall nor long-legged men, proceeded at our usual, reasonably-sedate pace when carrying out such a visual exercise in these climes: but soon enough, after just a mile or so, our sergeant was lagging somewhat behind. We carried on, only to see, when next looking back, that he was no longer even in sight. We waited until he came waddling back into view, then set off again. Only a short time later, in a wobbling kind of run, he finally caught up with us, and requested in heaving puffs and pants that we please proceed more slowly, as he wasn't used to going at our speed. We decided to stop for lunch not long after, out of respect for our unfit companion as much as anything else. We even shared our food with him, though he was supposed to provide his own. And then, come the remaining afternoon's trek, now in the real heat of the day, we had to take things at an even more leisurely pace, just for his sake. By the time that we rendezvoused back with the Jeep, he was exhausted to the point of collapse.

We did have to wonder what training the newly-formed Zairian army was giving to its recruits.

He got his revenge, though, when we came to catch the plane back from Bandundu to Kinshasa, and then onwards for home. The senior army officer manning the airport's customs post most unusually insisted in personally examining every single, smallest thing within our baggage, as well as carrying out a full body search.

Word had obviously spread.

****

Centrally landlocked beneath Zaire, Zambia provided not only my first taste of Africa, but was also where I carried out the last true management duties of my career. In the interim, it was a place which, in parts, I came to know well, paying many visits over a good few years, both as an agricultural engineer - including carrying out a not-insignificant national study of cane harvesting and transportation requirements on behalf of their government - and also as head of our management team at the Nakambala sugar plant. No great traumatic events ever happened to me there, I'm glad to say, and I enjoyed tremendous hospitality from both its staff and managing director.

There was also an ex-pat contractor, Will Lloyd, with whom I travelled to the USA on a number of occasions, and at whose family home I often stayed when in Lusaka, the nearby capital city. Will, a Yorkshireman by birth, had been employed on the infamous, ill-fated Tanganyika groundnut scheme in the immediate post-WW2 times; an exercise in the worst of agricultural project planning, execution, and management if ever there was one. It had, however, allowed him to develop a new system of tree-felling, which soon came into world-wide use. He'd then drifted over the border to Zambia and founded an earth-moving company, which soon became so large that he needed his own plane in order to visit all of his working sites.

It was that plane, along with its friendly pilot, which came to my aid when I once had to pay a scheduled monitoring visit to our team up at Kawambwa, in the very north-east of this near-bisected country. The drive up was intolerably tedious and tiring if undertaken wholly within Zambian territory. Its alternative, however, albeit much more efficient and practical by a literal long way, was to traverse the deeply-intrusive Zairian pedicle, a notorious area where armed gangs, often bands of irregular army units, robbed and abducted passers-by at will. We always tried to travel in a convoy of at least three vehicles when taking this route, keeping in close contact throughout the hour and a half's journey, and were much relieved to reach the border post at the end of the crossing, whichever way we were heading.

But this time, to my dismay, no-one else either wanted or needed to make the journey. I therefore faced the horribly-unappealing prospect of having, all alone, twice to undertake the marathon drive, or else to attempt the shorter round trip: if I ever even got as far as trying to come back, of course. As luck would have it, though, I happened to mention something about it whilst Willy was within earshot. He immediately offered to fly me up there himself. It was one area which he'd never visited, he explained. He'd be very interested, both for personal and business reasons, to have a quick look around.

And so, radioing ahead to notify them of our planned day and time of landing, off we subsequently flew. The small town's airfield, we soon discovered, was actually unusually large, its runway long and broad enough to take medium-sized jets: the reason, we strongly suspected, why it had really been constructed, so that sizeable military craft could easily be brought in should any 'local problems' cut off the area's main access via the pedicle route. It was also curiously empty; devoid, in fact, not just of any other signs of life, but of anything else resembling a plane. After a perfect touchdown, Will taxied to one end of the strip, and there we waited; and waited, and then waited some more.

No-one appeared for a very long time.

Eventually, an old pick-up truck very slowly approached from the perimeter road, drawing to a halt a good distance from our lonely little craft. Hesitantly, and certainly very cautiously, its single civilian occupant dismounted from his cab, and enquired - still from a good way off, and again, we thought, with some marked trepidation - who we were, and what we wanted. We explained, and then for good measure produced every bit of identification which we had with us, which seemed to satisfy him sufficiently. In fact, he suddenly seemed more than content, smiling warmly as he offered to drive us to the house where the "other Europes" were staying.

He was, so he told us, their equivalent of the local mayor. With no-one having passed on news of our intended arrival, he had bravely taken it upon himself to investigate and face up to what - or so the general local and severely-worried consensus had already concluded - was evidently the start of the long-feared invasion from Zaire.

Such, apparently, was the nature of life up there.

And life up near the border was definitely hard. There was already an established tea plantation some miles away from our team's base, but, apart from a little occasional work there, and the day to day catching and drying of fish, there was little else to occupy the population in any productive sense. Just to add to their problems, all supplies had to come in on the long, circuitous, and certainly costly route up from the 'copper belt' area to the west of the pedicle, as the Zairians either outright purloined any goods daring to attempt the direct crossing, or at the least demanded a very high customs tariff for them to be able both to enter and then leave again. Our current contact's remit, therefore, was to assess the possibilities of helping the locals to grow any type of food crops at all.

Our team weren't immune to those supply issues, either: and one time, just as in Nigeria, they found themselves running short of both food and fuel. It was coincident with one of my longer visits, and 'short of' fast became 'fully out'. Still, I at least came armed with a certain expertise. We sent one of our Land Rovers off on a desperate search and re-supply mission, and in the meantime curtailed our usual daily journeys and activities, conserved our stock of diesel fumes, and provided as best we could for ourselves. We'd heard from the locals that there was a nearby National Park which contained a good-sized river and some nice waterfalls. Equipped with fishing lines, small nets, and other appropriate paraphernalia, therefore, six of us piled into one small vehicle and proceeded in accordance with the directions given.

The green, grassy, tree-covered approach certainly appeared welcoming, if not very wet. But then, as we walked further towards the purported site, we began to hear a deep pounding noise, distant at first, then growing nearer and considerably louder - dare one say thunderous? - until we couldn't easily hear ourselves speak. Not, soon enough, that we had the words anyway: for before us, as we finally emerged into the open, lay the breathtaking sight of a whole wide horseshoe of huge, deep, frothing-white waterfalls, much like the famed Niagara or Victoria Falls, only apparently - according to some in our party who'd also visited the latter - slightly more impressive.

I did check later, and the Lumangwe Falls, on the Kalungwishi river, were indeed marked on a world atlas which I had back at home. But no-one to whom I've since mentioned them has ever had even the remotest idea of their existence, never mind how wonderful an afternoon one can spend dangling a few woefully-inadequate fishing lines over their edge.

We didn't, of course, actually catch any fish. We did pretty quickly lose all of the lines.

(As a note for the modern reader, these wonderful waterfalls are now easily to be found using your chosen search engine and mapping site. That modern font of all knowledge, Wikipedia, notes them as having a depth similar to that of the Victoria Falls - or, somewhat oddly, according to the Zambian Tourism website, slightly smaller - for which they are apparently often mistaken. Once a sacred spiritual site - oops, apologies for the fishing, then - they now form, along with two other nearby spectacular falls, a recognised national monument, officially opened to tourism in 1996. So there.)

****

Over on the other, eastern side of the continent, I made several trips into Ethiopia, where nothing very exciting happened on a personal level, but we were supposed to call everyone 'Comrade'.

We once visited to discuss a planned development which was to be aided by so-called 'financial backers' from Libya, the Ethiopian government having apparently refused to avail themselves of the World Bank funding already on offer. We didn't pursue this particular project further, and it turns out that the Libyans withdrew not long afterwards, too, leaving the Ethiopians high and dry.

It took several years before World Bank, obviously feeling a little piqued, were prepared to entertain assistance again; and one couldn't help but to observe that perhaps it served someone right.

****

For all sorts of complicated reasons which I now can't really remember, upon my one visit to neighbouring Somalia, the 'Horn of Africa', I had to adopt the guise of a German. We had a working relationship with a team from Essen who were going out at the same time, and so, with no need to do anything except just to mutter the odd occasional 'ja' or 'nein' as appropriate - a good job, as that was pretty much the full extent of my linguistic ability - we all flew out together from Frankfurt.

Landing at Mogadishu, already running late for our scheduled inaugural attendance at the Ministry of Reconstruction, we hurriedly filled in the required long and elaborate entry form. Everyone then passed rapidly through the remaining procedures; everyone but me, of course, to whom the immigration official for some reason chose to administer the full treatment.

Having gone through the entire contents of my cases, he then took my wallet and emptied it - cards, notes, scrap of paper, neatly-folded receipts and all - seizing with great delight upon some of the long, and now-very-old-type, signed credit card slips. These, he proclaimed, must be pre-approved cheques, or some other form of exchangeable currency - bringing money into the country was strictly forbidden, as it could be used to bribe people (!) - and promptly insisted that I also turn out my pockets: travel documents, tickets, coins, penknife, the lot. He continued to search for further damning evidence - of what, I'm not quite sure, but he certainly looked very hard for it, jabbering away in his local tongue whilst jumping up and down in a peculiarly-excited, yet, to me, quite threatening manner - until, nearly a whole hour later, one of his superiors finally arrived. He, fortunately, had a little more experience of Western money matters. What's more, he could speak a reasonable amount of English; enough, at least, to instruct me to pick and pack up everything whilst he signed my papers, and then escorted me to the door.

The entire German contingent had, of course, already rushed off: but their leader had thoughtfully sent his driver back to await my eventual appearance from the bowels of the airport, and I was thus conveyed to the doors of the ministry building just as the meeting proper, also including a large number of assorted other visiting international groups, was about to commence.

Somalia, in the early 1980s, was an even more 'Comradely' place than Ethiopia; the most communistic country, in fact, which I've ever entered, Russia included. It was also, now well over a decade after its formation from the merger of two former Italian and British colonies (not to mention more than a few other geopolitical complications), and then its subsequent independence, yet another country in a sadly-decrepit state. Everything was either utterly run down, or had plain already ground to a complete halt.

The factory whose viability I'd been sent out to assess, for instance, was not just hung, but positively smothered with cobwebs: the largest that I've ever seen, strung in-between old electric cables, girders, stairways, gangways, conveyor belts, and anywhere else which their creators - whose size I hated even to contemplate; although, given the general way of these things, was probably ever-so tiny - could possibly get to. We had to push and swim ourselves bodily through them, like some Halloween- or jungle-themed movie, just to get near enough to visually examine the machinery: which, pretty obviously, had all been sitting there, unused and untouched, for a very long time indeed.

The situation in the fields and with the irrigation systems was equally unkempt and uncared for. None of the agricultural implements or vehicles would ever be operable again. About the only interesting things actually worth looking at were the local herds of camels, used mainly for their milk, just like cows or goats elsewhere.

Yet for all that, it was still quite possible - in those days, at least - to wander around 'Mog' without feeling at all unsafe, even late at night. Moreover, the hotel was reasonably clean, the food palatable, and water still freely available 'on tap'.

Small mercies should never be under-appreciated.

Upon completing my inspection, I had to fly out via Rome. Disembarking from the plane, we were told that there was some kind of strike underway, so all the passengers had to retrieve their own baggage after it had been off-loaded and unceremoniously dumped upon the tarmac.

This, in my opinion, was not a bad thing in the least. It also brought me into close contact with a man from the World Bank's financing department. His team had been out in Somalia alongside us, and he was travelling home on the same flight. Shrugging and smiling, as you have to do in the circumstances, he addressed me with a number of words in German. I had no choice but to reply in English, much to his surprise: he could have sworn that I was a German, he said, as he'd heard me on many occasions talking 'shop' with my German colleagues.

So much for the odd ja or nein, then.

****

Mozambique, a good distance further down the eastern coast, was still - just - a Portuguese colony when I went there. It was also, clearly, right from the very start, a far from happy place. Government troops were everywhere, not just at the airport, but in the towns and villages, carrying out patrols in full wartime style along the bush roads, and generally looking suspiciously upon every single movement which the indigenous population made.

Brian McCloud and I paid our initial visit in order to evaluate an existing cane estate, Marromeu, situated on the south bank of the Zambesi river, not far from the coast. I was amazed - not to say extremely proud - to discover that their milling plant was one which I'd once designed and detailed, entirely on my own, all those years ago when I'd first worked for Fletchers. It was still working well enough, too.

We were actually staying at another estate, though: Luabo, a short distance downstream on the river's opposite, northern bank. The river itself was renowned for its ever-shifting bed, and whilst the large steamers which passed up and down had to maintain a state of high alert in order to navigate safely within the constantly-changing deep-water channels, we simply skimmed across the shoals on our daily commute in a small, water-jet-propelled boat.

It certainly made a nice change.

The river seemed to be a cultural demarcation line, too. Nearly all of the staff at Luabo were either British or Dutch, and the culture, entertainment, food, and so on were most definitely northern European. When we crossed the river, however, we found that the other estate had a distinctly Portuguese feel. Even the staff wives did exactly as artisans' wives would do back at home, sitting outside their doorways making bobbin lace. The only 'northerner' there, in fact, was the general manager, Ferdinand de Braacka. He, coincidentally, was the son of the first proper business contact whom I'd ever made overseas, in Mauritius, and I still remembered him as a boy.

Between that and the milling plant, I was starting to feel quite old.

Having reported our visit's findings, T&L were subsequently commissioned to carry out a full-scale viability study into Marromeu's expansion. In addition to Brian and myself, we then brought in a soil scientist, a factory engineer, and yet again Jim Villiers as our civil engineer.

This time, we all shared a bungalow in the Marromeu estate township itself. Brian, Jim and I usually went off every day in one Land Rover, and the remaining two took another, our trio sometimes travelling quite far from the estate to investigate an extensive grassy plain which we thought might be suitable for growing cane.

We were about half way across it one day, examining soil pits as we went, when we suffered a puncture. That, in itself, normally presented no great problems: but in this instance, there were two immediate additional aspects to be considered. Firstly, we couldn't find a jack in the vehicle: but, of course, we did have Jim's height and strength, so we weren't unduly worried. Secondarily, however, to our left, only a few hundred yards away, stood a large herd of grazing buffalo; to our right stood a similarly-sized, similarly-occupied, similarly-distanced herd of zebra. Each of these, if startled, could very quickly start an en-masse charge, likely to flatten anyone who stood - or pointlessly ran - in their way. Outdistancing them in a Land Rover was normally quite an easy business, given a short head start: but not when it had a completely flat tyre. Extreme caution was therefore required; as, once again, was a good engineering solution.

Quietly, and without making any sudden movements, we gathered up some large stones which were lying nearby. Then Jim grasped the rear of the vehicle and lifted it, whilst Brian and I piled the stones under its axle. We removed the offending wheel as if both it and the tyre iron were made of glass, and prayed that the stones wouldn't slip before we got its replacement in place. We then simply reversed the order, just as stealthily dismantled the stones, and finally, with all secure on board, restarted the engine. The animals all looked up, snorted, and shuffled around a bit, but seemed content enough to let us be on our way: which, being near to midday, was only to take us to the far end of the plain, where a shady grove of trees crowned the banks of a small river. It was a spot most eminently suitable for sitting upon whilst consuming sandwiches.

Rested and replenished, we carried on with the rest of our day's planned work, returning to base without further incident: something which, we later heard, hadn't been the case for an unfortunate, and soon to be very uncomfortable, local land overseer. Just a short while after us, apparently, he had also visited that exact same luncheon venue, for exactly that same purpose, although wandering further down to sit right by the water. Upon returning back up to his Land Rover, however, he'd found a large leopard stretched out along its warm canvas roof: sleepy, yet aware and wary enough of his presence that the poor man had been forced to wait, unmoving, amongst the trees until the animal, a very long time later, had finally decided to jump down and wander off somewhere else for the night.

We later learned that this area in general was favoured by the sugar company's own agronomist as a good place to go and watch the local wildlife. In his honour, therefore, we christened it Philippe Lefevre's Game Park, and marked our maps accordingly.

One Sunday, the three of us decided to drive the thirty or so miles north west, up along the river bank, to the town of Chupanga. There, we were told, lay Mary Livingstone's remains: and, indeed, we found a small, very British-looking church, its churchyard complete with many standing gravestones.

It didn't take very long to find hers, but, whilst wandering around, we couldn't fail to notice how many more were also inscribed with British-sounding men's and women's names. Most had only lived into their early twenties. They must, we assumed, have come out to staff the local mission which the Livingstones had founded, and one by one succumbed to the dreaded Black Water Fever, a malarial complication endemic in the area, given all the mosquitoes and flies.

Just before returning home, it was suggested that I should visit a land-clearing contractor up in the far north of the country, near the Zambia/Southern Rhodesia (soon to be Zimbabwe) borders. Their services, it was thought, might be of future use should the proposed Marromeu project get the go ahead, so it would be useful to see the work on which they were currently engaged.

This particular area was already well known for its considerable FRELIMO (the Front for the Liberation of Mozambique) guerrilla activity, not least because of their determined attempts to stop the ongoing construction of the Cabora (or Cahora) Bassa dam. Nevertheless, I was told that it was safe to visit, so I left Marromeu's little airstrip in their usual four-seater plane, and landed on a runway so wide that it wouldn't have done injustice to a major international hub. There, I'm glad to say, a pick-up and driver awaited me. But, not at all unlike my northern Zambian experience, there was absolutely nothing and no-one else about. The entire airport, I was told, had recently been abandoned by the Portuguese Airforce, being far too vulnerable to FRELIMO attack.

They weren't taking any chances on the road to the contractor's camp, either, our route almost constantly lined by numerous contingents of troops, all dressed in full battle gear, carrying out sweeps for the 'terrorists'. Still, I spent a very pleasant night alongside the now rather slimmer River Zambezi, and was able to have a good look around before my plane returned for me the next day.

I often think about what might have happened to those contractors, though. Not all that long after my visit, the independence fighters swept en masse through the entire northern area, brutally killing many of the Portuguese, as well as all of the local tribesmen who'd ever worked for them.

A couple of years later, FRELIMO forces arrived at Marromeu and Luabo, too. A pair of T&L engineers were still assisting at the estates. They were taken captive, and subsequently made to do all manner of menial tasks, including sweeping the streets. But that, fortunately, was the worst which befell them, and they were eventually allowed to return to England, mostly unharmed.

It was a stark reminder never to complain about my own minor problems and troubles. Not too much, anyway.

****

Sadly, I paid only one very short visit to the true North Africa. I'd been asked to advise if anything could be done to increase harvesting productivity within the Moroccan sugar industry.

Flying into Casablanca, I was then taken to the small resort of Mohammedia, consisting at that time of just one very modern hotel, a Moorish building adjacent to it, and a few native-style dwellings. I asked the hotel's staff what the older building was. It belonged to the president, they replied, and contained his harem. But, they continued, in rather hushed and conspiratorial tones, whilst prostitution was strictly forbidden in Morocco, the president liked to think that, with discretion, certain respected foreign visitors might still be able to enjoy the facilities which these premises offered. For a certain payment, of course.

Of course...

I quickly decided that, however discrete the visit, and even at the unspoken invitation of its president, I really ought not to disobey the law of the land. The hotel staff, frankly, seemed rather disappointed by my refusal. No doubt they were all in on the racket. I wondered if the president even knew.

Such matters aside, the field survey here was most interesting, if not ultimately successful. I'd never seen such an irrigation system before; nor since. The water which it carried came all the way down from the rains and snowmelt of the high Atlas mountains, finally distributed amongst the fields in a network of overhead channels, each made up of large, half-round drainage pipes, all held aloft on closely-spaced legs.

Unfortunately, the frequency of these supports left little room for any significant machinery and equipment to manoeuvre beneath. I made a few hopefully-helpful suggestions, but there wasn't really much - bar touting for the installation of an entire replacement water supply and management system; maybe even a nice new dam? - which I could in practice achieve.

Having swiftly wrapped up my primary purpose, I managed to fit in a brief visit to the beautiful old city of Rabat: and that, for me, made the whole trip worthwhile.

10

SWAZI SUCCESSES

My favourite part of Africa, I've left until last. Sadly, I never managed to reach the very southernmost Cape Province of South Africa, nor did I get to explore a great deal of that large country itself. But I did get to know one of its neighbours very well indeed: and it, again, despite its diminutive size, deserves a full chapter all to itself.

Early in 1975, Brian McCloud and I travelled out to prepare for a forthcoming full-scale feasibility study in the landlocked Kingdom of Swaziland (now Eswatini). A couple of years earlier, the Swazi government, desperate to enhance their national economic and export prospects, never mind to provide much-needed training and employment for the native population, had asked the Commonwealth Development Corporation to carry out preliminary investigations into the establishment of a new sugar estate and mill; the third in their tiny country. Our consultancy company had subsequently been commissioned to carry out a planning and development exercise, aiming to progress these ambitious, and very large-scale proposals.

We'd flown into Johannesburg, hired a car, and driven east towards the mountains which marked the kingdom's border. At the time, as a fuel-saving measure, the South Africans had imposed a fifty miles an hour speed limit; regular flashes from oncoming headlights helpfully indicated the many parked-up police cars and speed traps.

I'd been in the driver's seat for most of the long, sedate journey, but as soon as we crossed into Swaziland, Brian offered to take over. The main, single-carriageway road was of extremely good quality, and not exactly heaving with traffic. The natural assumption, being now out of the Republic, and with no indications to the contrary, was therefore that it could be travelled however one wished.

It wasn't long at all before a very large Swazi police officer pulled us over, immediately informing Brian that he'd been proceeding at a pace somewhat higher than currently allowed.

Brian, of course, protested most vigorously: there'd been absolutely no visible warning whatsoever of any further restrictions, once past the border post.

The officer surprisingly, at least in the strictest sense of the facts, agreed. The new Swazi limit had, he explained, been imposed only the previous weekend. They hadn't yet had time to put up a sign. But, of course, he continued, everyone already knew all about it.

Try as he might, Brian couldn't convince this staunch upholder of justice that such an all-encompassing situation might still not include two British businessmen who had just, only a mere few minutes beforehand, arrived for their very first time in his country. The mandatory fine of twenty Emalangeni was imposed, payment to be made at any police station within twenty four hours.

Fuming and cursing, Brian drove us onwards, dropping ever lower from the High to Middleveld lands, passing the country's small capital, Mbabane, and then its even-smaller second-largest town, Manzini. Here, about a third of the way to our destination, once past a small-scale commercial and industrial area, and then the 'international' airport, the main trans-country road abruptly changed from being a tarred to a wholly dirt one. It remained, however, both dry and well-levelled, and so, apart from being a little dusty, we arrived in otherwise good shape at the small town of Tshaneni, in the far north of the country's eastern Lowveld, close to both the South African and, a little further east, Mozambique borders.

The next morning's first order of business was to pay our fine. But Brian - being, despite his name, of Irish descent - was determined that he was going to get his own back in the process. Clutching a heavy bag full of the smallest coinage which the hotel's bar could muster - and were most pleased to get rid of - he thus stomped off to the local police station, still swearing liberally. He showed the desk sergeant the notice of fine, and then, with a flourish, poured out before him the large mound of coins, carefully pre-checked and counted to the exact sum due.

Yet instead of the anticipated fury registering upon the policeman's face, Brian was greeted with the happiest of smiles.

"Oh, thank you, Sir," beamed the sergeant. "Thank you so very, very much. I am completely out of small change, and was wondering where I could possibly acquire some this early in the morning."

Brian continued to make his feelings most patently clear for the entire rest of the day; and well beyond.

Over the following week, our team's number built up to eight, Jim Villiers again included. Headquartered as well as accommodated at the Impala Arms Hotel, we began touring the sizeable areas which we'd been asked to survey.

The Swazis' overall, far-from insignificant scheme encompassed not only the construction of a factory, along with all of its associated infrastructure, transport, and agricultural equipment, but a major new dam on the Black Mbuluzi river, providing irrigation for more than ten thousand hectares of cane. Two entire new townships were also to be built, in size ranking as the country's third and fourth largest, complete with shopping plazas, schools, police stations and post offices, churches, petrol stations, car parks, medical clinics, and social facilities for both the factory's staff and general labour. Additional provisions also had to be put in place to compensate and re-house the significant number of people who, unfortunately, had to be resettled as a result of the project's overall reach.

CDC, of course, were still involved. They already had considerable interests within the country anyway, including a substantial smallholder irrigation scheme ('SIS'), some paper milling operations, and the Tropical Management Training College at nearby Mananga, so were able to offer us a certain added entrée to others already holding good knowledge of the locality, and upon whose expertise we could therefore draw. And not a wine bottle was to be seen.

Access to all other manner of local farmers, economists, and the like came courtesy of the king's special representative, and our direct supervisor for the entire project, one Wagner Halasz: a most interesting character who, despite his non-Swazi, indeed wholly non-African heritage, acted as one of His Majesty's main and most trusted advisors.

Wagner was, in fact, Hungarian by birth. Escaping his homeland in the late 1930s, he'd landed firstly in Britain, and then immediately taken a boat out to South Africa, learning to speak English on the way. He then did the only thing which he knew anything at all about: selling seeds. This soon took him into Swaziland, where, with his enthusiastic, energetic, and constantly-active ways, he quickly established a reputation for being able to obtain all manner of things. It wasn't long before his reach extended right into the country's highest circles.

Technically, although Wagner now spent most of his weekdays in Swaziland, he still lived on the outskirts of Jo'burg, from where he also ran several businesses, currently managed by his very efficient secretary, Mrs MacDonald. He clearly enjoyed being busy, though, for on top of all of his other interests, and never mind the demands of his many state official and advisory duties, he was also the designated 'entrepreneur' in charge of sourcing both uniforms and weaponry for Swaziland's army. How official this particular role really was I don't know, but we ended up sharing an office for a while, and not only did we often share confidences, but sometimes I had to take his telephone calls. It was a little difficult to get over to certain callers, whose identities I didn't even want to know, never mind write down, that whilst I was certainly the man interested in supplies of trucks, tractors, and sugar milling equipment, mortars and machine guns really weren't my thing at all.

Wagner also had a fascinating circle of friends, many of whom I came to know. Roger Feltzman, for instance, in due course becoming the new sugar company's solicitor, was, like so many of Wagner's close contacts, a Jew. He was also, at social gatherings, the best teller of Jewish jokes that I've ever met. Another acquaintance, and another Hungarian, ran a very good restaurant near to Manzini; a third, Kurt Hennet, owned a very successful fruit farm at Tambankula, where he also grew cane. Kurt, I was later told, had been sent out by the Hungarians at the end of the Second World War with orders to find lands on which to settle European Jews. Landing in South Africa, he'd slowly trekked, along with his wagons and horses, deep into Swaziland's then undeveloped and oft-uninhabited lands. He never fulfilled his formal mission: but he did settle himself, along with his wife, who'd already fled to England in the pre-war period. It was Kurt, in fact, who was eventually credited with planting the metaphorical seeds of the whole new sugar estate idea, having once suggested to royal ears that there was significant scope to increase the growth of cane and the allied production of sugar.

The Kingdom of Swaziland, the last of its kind in Africa, is, in today's terminology, an absolute hereditary monarchy. His (now late) Majesty King Sobhuza II KBE, 'iNgwenyama', was, both then and to this day, officially the world's longest-reigning monarch, having acquired his role at the tender age of four months old, and eventually holding it for nearly eighty three years. During that time, he had taken almost as many wives, and produced children numbering well into the hundreds.

Whilst certainly not averse to encouraging modern methods and developments when they benefited his peoples, the ageing, mostly-kindly king also continued to rule his country in the traditional manner, retaining all the tribal chiefs - and the odd outsider - as his advisors, but then, having convened his Council and heard each one put their case or render their opinion, reaching and decreeing all subsequent final, and thus unquestionable, decisions alone.

Or maybe not so much, for, as I understood it, the traditional counterbalancing influence of an older or senior female relative was also always required; an actual role, function and process which, to be honest, I never quite got the hang of, but doubtless will have been, and still will be, seen by some as a jolly good idea.

As far as I also heard things, it was actually, if unwittingly, the CDC, through their SIS scheme, who'd brought this old system of rule back into being. I say 'back' because, in the early 1960s, the British government, the country's former colonial rulers, had wanted to modernise matters. Consequently, they'd proposed a form of constitutional self-government, forming a 'house' structure along UK parliamentary lines, and even presenting a ceremonial mace - a full replica of that used in Westminster - for its 'speaker'. The first elections were held just prior to independence being granted in 1968. The SIS project, however, was by then also developing into a wider form of democratic land reform. Every farmer within their scheme held an equal portion of irrigated land, became part of a co-operative sales organisation, was entitled to equal voting rights at their meetings, and, in fact, became, in relation to most of the population, very wealthy and privileged indeed. When the new parliament was formally brought into being, most of the rest of the country sent representatives of their tribes: under the general approval and oversight of the king, of course, and thereby becoming known as the 'king's men'. The members sent from the SIS areas, however, instead considered themselves to be 'new men', with new political aims and ideals, and were determined not to be subjugated to, or by, those of the 'king's party'.

Inevitably, it wasn't long before the 'new party' demanded a vote on some proceedings or other, and duly went into the 'no' lobby, against the tribal 'yes' men. The king wasn't going to tolerate these upstarts, these proud, newly-entitled, newly-wealthy landholders upsetting his kingdom's traditional ways, nor taking away an undue amount of whatever power he still retained from under him: and so, about two years before our arrival, he had promptly found a way to dissolve the parliament, reassumed all of his former control as head of state, and simply carried on just as he'd done before. There was nothing which SIS could at that point do, except to accept the situation, and go on making its farmer tenants richer and richer; until, at least, the scale of industrialisation within the country gradually increased, and their influence accordingly waned.

The governance situation did also, to some degree, revert back to the planned constitutional system after a few years, and elections of a kind were, I recall, again being held just as we commenced the main part of our project work. Yet the monarch still, to my knowledge, retains a good degree of supervision, if not actual control, over the appointment of many of the members of both parliamentary houses. And although undoubtedly suffering its ongoing problems, perhaps the continued influences and acceptance of this tribal monarchy did spare the kingdom a good deal of the effects, troubles, and various ongoing political trials and tribulations which I experienced in so many other countries in their own post-independence days.

Politics and rulers aside, or even notwithstanding, Swaziland, despite its most definitely 'developing' status, and the very basic standards of living and undoubted poverty of its people as we would have judged it, was a most pleasant - and certainly very beautiful - place in which to work, rest and play. We didn't lack for social and leisure opportunities on this trip, either. The SIS scheme, the Mananga College, and the existing Mhlume sugar mill were all close enough to our Tshaneni base, and their staff all freely shared the respective facilities, together with those at the official Yacht Club - as, of course, you'd expect in a tiny, land-locked African country - situated at the nearby Sand River reservoir. We T&L men, besides frequently contacting their employers regarding our survey, were often invited to join them.

Towards the end of our initial visit, we received three separate requests to dine with the senior representatives of each of these aforementioned institutions; all to take place within the same week. Dress for such occasions was rarely specified, and, consequentially, it was always difficult to know what to wear (not just a problem for the ladies!). For the first evening, at SIS, we elected to attend sporting our standard full business suits. We had, of course, completely overdone it, and were immediately ordered to strip back to more comfortable open-necked shirts. This, we therefore decided, was obviously the accepted norm for the area; and so, two nights later, when we went off to Mananga College, we left our jackets and ties back at the hotel. To our embarrassment, as soon as we entered the building, there stood our esteemed host and his other guests, much more formally attired. Fast forward two more days, and we thought it best to revert back to jacket and tie: only to be told upon arrival at Mhlume that we shouldn't be such stupid, stuffy Brits.

Lesson: you can't get everything right all of the time.

Perhaps because of its oft-visiting South African neighbours, or perhaps because of its past British colonial influences, or perhaps for some completely different reason - but certainly not mass, world-wide tourism, back in those days - Swaziland also already had more than its expected fair share of fine hotels. These were often modern in their construction, facilities and layout, yet usually retained a certain 'native charm' in their design and decor, many incorporating large thatched verandas and outdoor dining terraces, swimming pools surrounded by blossom-laden Bourgenvillia bushes and Birds of Paradise flowers, and small, grass- or palm-roofed 'rondavel'-style huts, tables and chairs.

In June 1975, having completed our investigations, we were asked to present their outline findings to a gathering of various VIPs and other generally-interested officials. The meeting was scheduled for early morning, at a government office in Mbabane. Wanting to arrive not only in good time, but also not covered in dust from the long road journey, we travelled up from Tshaneni the night before, booking into a previously-unknown hotel just outside the mini-capital.

The Highland View, for once, actually lived up to its name, looking down over the huge Ezulwini Valley. We arrived late in the afternoon, checked into our rooms, and, after an early evening meal, congregated in a corner of the lounge bar, next to a window which afforded a magnificent view of the lush, green, often tree-covered hills as they sloped and rolled their way down through this 'place of heaven' towards Manzini.

We could also see down the full length of the room. Soon enough - although none of us actually mentioned anything to each other at the time - we'd all noticed a particularly attractive local girl lounging against the bar itself. There was also, we noted idly, a sizeable gathering of several rather plainer, sometimes older women towards one of the faraway corners. The connection, and then pattern, quickly became clear as, from time to time, one by one, men began to come up and chat with the lovely young lady. After a short while, their not-so-innocuous conversation ended, she would slide off her stool and accompany the man through a side door which led to the outside yard. At the same time, one of the 'not-so-decorative' group would also detach herself from the others, and swiftly follow the departing couple out into the darkness. Within a relatively short timeframe, the first girl would then re-appear, and the whole process began again, repeated over and over across the course of the evening until, finally, there were no more 'less attractive' ones left. It was obviously some form of set-up for the local 'ladies of the night': but we could never quite work out exactly how it operated, nor whether the men actually got to choose, or get a good look at, or even knew, who they had with them before they'd committed to, as it were, sampling the goods.

Whatever the true answer was, whoever was actually organising and benefiting from it, we didn't care to investigate further, and the proceedings certainly didn't form any part of our report of the following day. The king surely knew all about it, anyway.

Not too far below, close to Lobamba - both the historical administrative capital and location of the Royal Village compound - sat the Royal Swazi Spa. This hotel also contained a grand casino, which attracted many white visitors from the Republic of South Africa, often driving or flying down just for the weekend. Not only did they come to gamble, prohibited in their own country, but to meet with their 'coloured' girlfriends and mistresses, whose company was, at that time, not permitted back home.

The more genteel Swazi Inn, on the other hand, had been converted from a very old thatched building, and was well-known for its excellent food and fine dining. It had once been the dwelling of one J. Rider Haggard, and was purported to be where he'd been inspired to write, and had then duly written, *King Solomon's Mines.* It also possessed a very fine view, the main building's large picture windows looking directly onto the high twin peaks of a mountain which lay well the other side of the long, wide valley, and was visible from far around; as soon as you entered the country through the Highveld, in fact. These rocky promontories were reputed to be the location of those aforesaid legendary mines, and were known, given their connection to said same ancient king, as Sheba's Breasts. It was a perfectly-acceptable name for a culture whose womenfolk still habitually, if now much less publicly, dressed topless.

It was also a source of some considerable amusement and embarrassed giggles - should one say titillation? - as regards my English wife and teenage daughter when they came out to visit during the summer holidays, a rare luxury allowed to me whilst on the Swazi project. We spent a few days together up in Jo'burg upon their arrival, and they then accompanied me on a brief business trip to Durban, further down the South African coast, where we stayed next to - sadly not in - a very fancy, modern, beach-front hotel whose frontage sported two extremely-novel external glass elevators, and overlooked an open-air snake park: which, yes, we did very cautiously visit. Then we travelled back through both the famous Kruger, and then the several much smaller - but actually equally-satisfying, especially if you like counting breeds of antelope - Swazi game parks, and enjoyed several oil-drum 'braais' (barbecues) along with my colleagues and their families over on the reservoir's shores.

The native women used to make and sell all kinds of jewellery, but especially necklaces, their shapes and designs constructed from threaded strands of tiny glass beads. As it happens, Rai had just started making her own bead jewellery back in the UK, albeit using a small, tabletop loom. It was to her mixed bemusement and embarrassment, therefore, that, whilst wandering through Mbabane's main outdoor market, the traditionally-dressed 'bead ladies' kept beckoning her over vigorously and reaching out their hands to examine with fascinated interest the hand-made choker which she then wore. They couldn't seem to work out how she'd made it; or perhaps they wanted to know where she'd got it from, and what its imbedded patterns represented. Sadly, with a large language barrier, they never got to be told.

Lest you gain the wrong idea, by the way, we, the working men, certainly weren't spending all of our time lounging around in restaurants, bars, casinos, and posh hotels; not even social clubs, by the time of our next return.

The huge, complex project had formally gained the go ahead a year or so after our initial report: December 1977, the official factory inauguration pamphlet tells me. Contracts to manage both the scheme's construction and general programme phases had, in their entirety, been awarded to the two T&L subsidiaries with which I was most closely associated. The works also encompassed all of the irrigation canals and water distribution infrastructure from the foot of the dam onwards, all ongoing, post-completion operations, and the concurrent provision of sufficient education and training to enable Swazi nationals to achieve full management competence of this huge, complex project in due course. There was now, therefore, even just in dealing with the initial start-up requirements, in both field and office, in Swaziland, the UK, and elsewhere, a great deal of work to be done.

We got started at once. By late 1978, if not mostly a lot earlier, T&L had put together, on behalf of the Swazi government, a consortium of some twelve or more world-wide financing, investment, and industrial groups to fund the entire development, including the new dam, and handle all of the other requisite technical, commercial, financial, and legal aspects; and anything else which might come up.

The one thing which hadn't by then come up, though, was the development's actual name. Always simply referred to by everyone, no matter how intimately or officially connected, as 'the third sugar mill', it had long been decreed that there was to be no other name until the king finally gave it one. No-one knew exactly why this was, nor if His Majesty even ever had a name in mind, nor when or why he actually came up with it: and the most closely-guarded secret it remained, even from Wagner. But finally, in October of that year, come the appointed day for its official dedication and ground-breaking ceremony - construction, as usually happens, having from our side officially started some months earlier - it was proudly pronounced that henceforth the project would be known as the Simunye Sugar Estate. Simunye, in siSwati (or Swazi), means 'we are one', and whilst I don't know exactly what the king intended to convey to his subjects and others, it certainly seemed to me to be a very appropriate name for a scheme having so many collaborating participants.

My immediate boss and managing director, Stephen Gaine, was placed in overall charge of the whole development phase. I, as well as acting as his deputy, was appointed its general co-ordinator, liaising, amongst many other things, between our Durban-based factory design team and the two site-based design and construction teams who were handling the respective factory and civil works. Through my own specialist department, I also undertook the specification and design of all the required transport elements, including personal, agricultural, and mechanical harvesting equipment, and worked alongside the factory team to design and obtain all of their unloading and cane-yard machinery.

In addition to my various other duties, as if they weren't enough already, I was then asked to act as the first general manager for the Royal Swaziland Sugar Corporation, newly-formed and formally registered, but not as yet having any physical, never mind practical form, nor proper operational function.

All in all, therefore, between my trips to Swaziland and neighbouring South Africa - and never forgetting my other pre-scheduled inspections and supervisory visits, to Trinidad and Zambia in particular - I was being kept pretty busy.

My chosen base of operations this time was near Lobamba, in the Ezulwini Valley's Holiday Inn. As well as being a pleasant, modern, and not overly costly hotel, it was about as central as you could get between the new factory's location - south of our original survey base, but still well to the country's far north-east - and Mbabane, where I was currently sharing a small office room with Wagner Halasz. But it was now time to seek out larger facilities to accommodate the company's initial staff.

I'd discovered that there were two vacant wings in some buildings originally built for a television set construction company (of all things!) which had recently relocated into the RSA. They were very usefully situated near to the newly-developed Matsapha airport, but, upon inspection, the space on offer also proved to be rather larger than I required. As it so happened, though, I fell into idle conversation with a fellow hotel guest a few days later, and it transpired that he was busy setting up the new Swazi national airline, and also looking for some suitable premises. Between us, therefore, we took on a shared lease, our half being passed on to the airline once permanent site offices became ready late in 1979.

My new temporary fifedom - my very own, to start with, as I was the only one there - contained a central reception area and five individual rooms. I allocated one room for the purchasing department, one for accounts, one for agriculture, one to general clerical and typing, and the other for the general manager - me, and then whoever came out as my permanent successor - and hurriedly furnished them all with desks, chairs, and cabinets. Having eventually located somewhere to buy all of the necessary ledgers and other assorted stationery, I hastily designed the layout and arranged the printing-up of sufficient order forms, letter heads, business cards, and everything else needed to initiate the general operations of our soon-to-be large, international company. A competent typist was quickly engaged, and, by sharing a receptionist with the fledgling airline, RSSC's operations metaphorically also got off the ground.

I wore, to begin with, a great many hats, flitting back and forth between one empty room and another with carefree abandon. With a filing system organised in the purchasing office, I drew up orders for the first stage machinery and transport. Then, popping next door to the accounts office, I entered up the ledgers, and thus commenced the company's formal financial records. In the agricultural office, meanwhile, my own equipment specifications were prepared, and contracts were placed for the initial land clearing and soil preparation works. In-between, any in-person callers were interviewed, and numerous telephone conversations held with all kinds of existing and prospective contacts, whilst living it up in the GM's personal quarters.

I'd by now taken delivery of six executive cars and a similar number of pickups; transport, as and when it was needed, certainly wasn't a current issue. Doing all of the leg work was, however, becoming quite a big one: so Wagner - who'd by now added the title and duties of company chairman to his portfolio - introduced me to Anthony, a young local man who'd just finished college and was seeking commercial experience. He got it, becoming RSSC's very first clerk; its very first proper employee, in fact.

The land development contractor moved onto site in January. I pretty much immediately realised that I'd need an agricultural assistant, just to oversee their work. Fortunately, I was able to secure the services of Phil David, a man whom I already knew well, and he was rapidly on scene. Others then trickled in: a company secretary from T&L's Durban office, and Michael Clef, the company's chief accountant from the UK. Vinny Groul was appointed as site project manager, although he didn't need to take up space in my offices, and a permanent overall company general manager, Graham Guide, was eventually nominated to head everything up. Graham, however, was coming in from our Zambian estate, and couldn't be released until late April, so I continued in his stead until then. Not, of course, that I was in the office as much as I was out, still visiting the Durban design office frequently, not to mention the multitude of various prospective suppliers and contractors to be seen up in Jo'burg, and with several repeat calls to be paid on our neighbouring cane estates, never mind the other two existing Swazi mills, and then... Did I mention that it was a busy time?

Only after Graham's eventual arrival, and a subsequent week of overlap and briefing, was I finally able to relinquish control and, with a great sigh of relief, return to England for a well-earned rest.

Come mid-1979, the main wall of the new dam had been completed, and Lake Mnjoli was, if slowly, born, its name - again chosen by the king - apparently, and rather appropriately, referencing their god of rain. Once fully filled, the resulting irrigation system was scheduled to become fully operational early the following year, together with another Umbuluzi river pumping station and a substantial weir. The factory site had also been cleared of its scrub, bush, and trees, and was now level. Much of the required equipment and materials was already in place, merely awaiting final assembly. The allocated growing lands were starting to be prepared, and many of the houses in one of the planned townships, Lusothi, were not only built but occupied.

The project's official primary phase - the first canes planted and irrigated, the erection and commissioning of the new factory mill, together with its offices and ancillary buildings, and the construction of many houses and accompanying social facilities - was deemed complete early in 1980. By now, I was in charge of ongoing site development, and as the first commercial trial crop was taken off around that time - the first cane being officially crushed by the newly-operating mill in May of that year, a month ahead of schedule - I can therefore also claim to have been its first - if again only acting - estate manager, pending a more permanent, full-time occupant for the post.

Although technically already commissioned and operational for some months, the king formally inaugurated the factory and estate at the end of August. The Financial Times in the UK even did a full

4-page pull-out section to mark the occasion. Ellie was invited to fly out to join me at the ceremony - at which we were proudly seated in positions of great honour, right in the front row - along with the preceding night's Royal Celebration Dinner at the Royal Spa Hotel: to her delight, comprising a most contemporary English-style menu of melon, some kind of eel-like fish, roast lamb with croquettes, carrots, beans and cauliflower, then Swiss roll to finish.

More importantly, it afforded us a few precious days together, which we decided to spend lazily touring the local countryside. My hire car, however, had other ideas, its engine seizing up towards the end of our first day's drive along the back bush roads. Still, at least this time it didn't happen anywhere, or in any manner, approaching as exciting as, nor requiring any form of daring action or rescue even near approaching, any of my former exploits; it, in fact, had the full and common, courteous decency to fail only as we were right about to rejoin the main road, so we were soon recovered, repaired, and back on our little tourist trail.

The Gods must finally have turned to my side!

I continued to make regular and frequent post-construction visits out to Simunye in my supervisory and management roles, making sure that the harvesting and transport systems were working according to plan, and that the other, later phases of the works also went to schedule. (That, and the witnessing of several world 12-hour mechanised cane-cutting records falling amidst the friendly, yet seriously-competitive rivalry of 'our' and the neighbouring sugar estates; a record which, to my delight, Simunye won from Mhlume in July 1982, by a very impressive margin.) Within three years, about half of the eventual planned cane lands had been planted up and irrigated, and I needed only to attend once every twelve or so months.

It was on the very last of these visits - actually, one of my last ever overseas trips - that I sustained the only serious injury of my whole working life. I was doing nothing more strenuous or dangerous than simply - and quite adequately soberly - leaving the Simunye Social

Club after enjoying an evening meal. It was pouring down with rain, so I ran straight back to my nearby bungalow: and promptly slipped and fell headlong onto its wire boundary fence. The taut top strand sliced deep into my upper lip - it, from previous experience, already seeming to have been designated as the target for any lasting damage ever to be inflicted upon me - and, for the first time in all of my overseas stays, I had to seek the urgent attentions of the company doctor. Very quickly, although it was quite late at night, he had me in the local medical centre's theatre, where twelve neat stitches were soon holding everything together again.

I'd had considerable influence in designing that clinic, taking advice from several of the UK's top corporate doctors as to what might be required: but I'd never expected to have to take advantage of its services personally. Nevertheless, I was pleased to know how professional and efficient it had turned out to be.

I'd also had some input into the social club - well, frankly, I'd had quite a lot of input into quite a lot of things, one way or another - and it was they who decided to hold a dinner in my honour before I departed for home. This was to "celebrate" my retirement, the invitation said. I hoped that was meant in the way in which it was taken. The meal was, so I'm told, lovely - for those who could move their swollen lips adequately enough to take it all in - and the evening was certainly memorable: even if the manager did type up an annotated, nay narrative menu which contained more than a few passing references to people falling at fences, laughing until they were in stitches, making home runs, and all the like.

A few days later, I departed Swaziland for the final time, passing briefly through Zambia on my way home. After two weeks renewing many friends and contacts, and although not quite yet scheduled to be put out to pasture, Don Silver, the then resident managing director, made a formal presentation to me as a memento of my long association with Tate & Lyle's estates.

It was, sadly, soon to be a final departure for Wagner Halasz, too. Six months later, amongst the many wonderful letters which I received from my former Simunye colleagues around the time of my official retirement, came also the news of his sudden, unexpected death. But he went out in reasonable style, they said, suffering a massive haemorrhage in the midst of dinner with some senior government minister. His loss was certainly felt by the company, never mind the country. I couldn't get over in time, but his funeral service in Mbabane was extremely well attended, by expats and Swazis alike, with eulogies and thanks being offered not only by senior representatives of T&L and RSSC, but from a grateful adopted nation.

Almost immediately afterwards, the sugar estate was hit by a significant cyclone - its second, after a major 'once in a hundred years' storm only the year before - bringing such heavy rainfall that several of the irrigation canals and bridges were taken out of action, along with roads and other infrastructure works. Fortunately, however, despite "a few anxious moments", as Graham Guide subsequently wrote to me, any more major flood damage was this time averted.

I'm only grateful that I didn't have to deal with the aftermath of either of those two events. But by then, I had said goodbye to Africa for ever.

11

CENTRAL AMERICAN SAMBAS

At the base of the Central American mainland lies a country which, although probably sounding rather strange now, ranked amongst my overall favourites: Colombia. For the majority of my time there, little to nothing was known of the problems for which it later became so renowned, the only significant pitfall awaiting its average foreign visitor in those days being the purchase of plentiful, cheap, and extremely-illicit emeralds.

My first trip out was to the Risaralda Valley, a central, inland area set between several ranges of lush green mountains. It was then best known, in the early 1970s, for producing some of the world's finest coffees. Our client, the 'Caffeteros' (a national organisation of coffee growers) wanted to know if they could also grow sugar cane there; maybe establish a factory, too.

We'd brought together a team from various sources for this task, headed by John Hartley, a former general manager of Caroni in Trinidad. He was a man whom I didn't know at all well back then, as he'd always seemed so terribly aloof when encountered in his previous, exalted position. As it was, when I subsequently got to know him and his wife rather better, I actually discovered that John was perfectly and very pleasantly friendly, just remarkably shy.

The others had already been at site for about two weeks when I arrived. An easy forty-five minute flight from the capital, Bogota, took me to our Pereira base, where I was met not by John - one way or another, not entirely unexpected - but by someone whom I'd never encountered before, the accountant engaged from a private firm to handle the survey's financial aspects. Pretty much his first words to me, after the usual preliminary greetings, were to the effect that something was "very wrong with Hartley", that perhaps he was ill, but in any event he wasn't currently in any fit state to direct the project's affairs.

Wonderful news!

The next morning, at my first meeting with the rest of the team, I was promptly told that, as the senior T&L man present, I must make the apparently-awkward decision about what should be done. So I went to see John in his 'sickroom'. There was no urgent medical or surgical need, as far as I could tell; but it certainly wasn't hard to see that there was a clear problem. Quite suddenly, and without any apparent reason or warning, he'd simply lost absolutely all confidence in himself and his capabilities, and was now, in my layman's estimation, in the full throws of what we then termed a 'nervous breakdown'. I hastily arranged for him to be flown straight back to the UK, accompanied by another team member who'd more or less completed his work, and took over the study's supervision myself. (Incidentally, whilst in time recovering well enough, John never served overseas again. He did, however, write a very good 'sugar handbook', part of which I edited when his wife eventually published it after his death.)

The rest of my visit - obviously more lengthy than originally planned - was highly enjoyable, and incorporated several interesting, otherwise unscheduled opportunities to visit the Cauca Valley, down on the Pacific coast, then a very well-developed, wealthy, and prospering place which produced the majority of the country's raw sugar, as well as other crops.

The three main coffee farmers with whom we regularly met seemed determined only to make my stay even more pleasant. Not only did they provide us with efficient, English-speaking secretaries to type up our findings into a preliminary report and then final presentation, but, with fine hospitality, they introduced us on both the social and business levels to many of the local people, their own families included. One later invited me to spend a weekend up on his mountainside coffee estate, an experience which gave me a whole new insight into this major Colombian industry. Before we left for home, they also insisted on taking me and Gerald, the aforementioned accountant, to see the Gold Museum, where many ancient Aztec and Inca artefacts were stored; and a place, apparently, to which only a very few select visitors were ever allowed to go.

The Caffeteros had obviously developed a taste for the sweet stuff: we were invited back a couple of years later by another group, this time for a feasibility exercise up in the north-east.

The small city of Cucuta was served by its own airport, with regular flights to and from Bogota, and had all of the facilities which one might reasonably require, including plenty of shops and restaurants; even a sizeable night club, somewhere on the outskirts. It also adjoined the Venezuelan boundary, a stony-shored river forming the two countries' border for a good distance both north and south, and a nearby bridge providing not only the official international entry and exit point, but actually, as we later discovered, the only dry means of crossing the river at all.

There were, however, many unofficial crossing points, albeit mostly able to be attempted only at times of lowest flow; times which usually coincided with the cane-harvesting season. Many of the Colombian sugar farmers therefore habitually used them to access their nearest processing factory, which was just a few miles away, but over on the Venezuelan side. These 'fords', of course, offered a great time-saving convenience: but they also most advantageously relieved the growers of significant taxes being levied by the Colombian government on every load of officially-exported cane. The Venezuelans were understandably more than happy to turn a blind eye: cane supplies from their country alone were nowhere near enough to make the factory financially viable, and they were quite willing to pay the same amount per ton of cane delivered, wherever its origin. It was a win-win all round; except for the Colombian Treasury.

Our team comprised, initially, a group of four from T&L's Bromley office: an agronomist, an accountant, a civil engineer, and me as both transport engineer and overall leader. A fifth man, a factory engineer, joined us later. We were provided with accommodation within three adjacent bungalows, part of an hotel complex set besides the road leading out towards the river bridge, plus two good four-wheel drive vehicles, an office in one of the outlying areas of town, and a daily monetary food allowance which enabled us to eat out wherever we chose. We were also assigned the services of a local government agriculturalist and his wife, not only to act as guides, but to introduce us around the various local farmers and other appropriate authorities. Both Ingenio and Felice spoke excellent English, and not only did they act in their commercial guises, but soon became our constant 'out of office hours' companions, too.

Our objective was to find out if sufficient cane could be grown on the Colombian side of the river to supply a new factory of their own, and thus avoid having either to import processed sugar from 'over the river', or to transport it all the way up from the Cauca, and soon-to-be Risaralda, mills. So we proceeded to explore the area, including gaining permission to look around the Venezuelan factory, looking for suitable land. Soon enough, we were extending our search further afield, to the river's east, where Colombian territory stretched along both banks. But, whilst we could plainly see it, we simply couldn't work out how to get there. Driving at least thirty miles downstream of the city, there still wasn't another bridge to be found.

"Not to worry," announced Ingenio, at our lead car's wheel. "I'll show you how the locals do it."

We assumed that he'd take us to another hidden stretch of shallow water. Yet, as our two vehicles drove further along the main western bank's road, the river instead appeared to become increasingly and consistently deeper. And then Ingenio suddenly turned off onto a surprisingly well-surfaced side track, and we found ourselves facing not only the river, but a white-watered weir. Spanning the entire river, the 300 or more feet between its solid abutments was connected on the upstream side by a vertical concrete wall, about six or seven feet high, over which fell a constant, fast-flowing curtain of water, landing onto another grey cast apron - or, more correctly, cill - of perhaps twelve feet in width. This, in turn, sloped gently downwards before terminating abruptly in a good two-foot drop above a broad stretch of large stones, deliberately placed to break up the overspill's force. The water cascading down the cill was currently about a foot in depth, we estimated: and Ingenio continued to drive straight into and then on through it, splashing his way right over to the other side, where another hard-surfaced track could be seen rising up to the main road atop the other bank.

We, in the second vehicle, just gulped: yet there was nothing else to do but follow. The descending water was constantly pushing us sideways across and down the slippery, sideways-angled ramp, no matter how much we fought it. But fight it we did, soon realising that the secret was to compensate by steering straight towards the 'waterfall', and never, ever, try to stop.

Shortly thereafter, we also arrived back onto dry, and reassuringly-unmoving land. Only then did we notice the official-looking signs at each side of the weir, both stating, by reference to a depth gauge set below the wall, when it was respectively safe to cross in a car, a high four-wheel drive vehicle, and a truck. So this was an approved route after all: or, rather, as was later explained, it was actually a structure originally erected to control irrigation water in the river's upper reaches, but then, in the absence of any other form of local crossing, adapted by its own engineers into an albeit worryingly-narrow, but clearly constantly-used, roadway as well.

There were, inevitably, many intrepid, or foolhardy, souls who were quite used to venturing over when the water level was much higher than considered safe; and, inevitably, sooner or later, some were washed over the 'rip-rap'. Most suffered nothing but a severely bashed and battered vehicle, accompanied by a most ignominious, difficult, and definitely soggy pedestrian exit from the shallows' sides. A few, however, did occasionally get swept out into the much deeper water beyond, with consequentially dire results.

We must subsequently have used that particular means of getting across the river many, many times: but, probably thankfully, we never even got to see it, never mind attempt it, at anywhere near full spate.

Our day's work over, we'd often gather upon one of our bungalows' balconies to sample both the local beer and the excellent local rum, distilled not too far away, before deciding where to eat and spend the rest of the evening. The latter was often quite an international affair, as, just a little way down the road, and over the Cucuta bridge, there was a very good Venezuelan restaurant. Here, amongst many other traditional delicacies, they served 'parrilla', portions of various meats which you DIY-cooked on small barbecue-type fires which the waiters placed next to your table. They also served a very palatable pigeon stew: although I wasn't so keen after one night finding that it included the whole pigeons' heads, too.

Felice and Ingenio also took us to various local restaurant-cum-dancehalls which they particularly liked. The pair were very fond of dancing, and would often whirl and wiggle their way into the wee small hours, whilst we less-inclined types, with the transport to get both them and us home again, simply waited around. On one occasion, though, Felice declared that we should all accompany her to the 'proper' night club, out at the edges of town. It wasn't exactly our preferred choice of venue, but she was always so kind and helpful that it seemed rude to say no.

The club was spacious, dimly lit, and filled with bright, constantly-moving spotlights; and, of course, very crowded. We managed to find a table around which all seven of us could just about sit in comfort, and ordered our drinks. No sooner had these been served than a group of girls in their late teens came sauntering over, trying to persuade us to let them join our table, to dance with them, and, of course, to buy them drinks. This was nothing but a usual management ploy which we'd all encountered too many times before, so we simply pretended that we didn't understand. The girls, however, had already recognised our Columbian friends, and proceeded to ask them in puzzled, disappointed tones why their guests didn't want to get to know them better. Ingenio reassured us that the venue wasn't used for prostitution: although, as the evening progressed, we did notice several young ladies very quickly linking up with, and then disappearing off the premises arm-in arm with, various assorted men, so, at the very least, it definitely served as some form of pick-up joint. Still, we made it clear that we weren't interested in any additional company, activities, or knowledge of whatever kind, and eventually the girls left us in peace.

Well, relative peace: the club's music was, to say the least, rather loud, and our table was set so close to a huge stack of speakers that we couldn't talk properly to each other at all. Not, at least, until our factory engineer, Nick, sitting nearest the wall, spotted where the wires led into the back of the cabinets. By the simple expedient of reaching down and applying the requisite sharp tug, he was able to reduce the sound levels in our vicinity considerably. No-one else seemed to notice, and the rest of the night passed in much more pleasant conversation.

Until, that was, it came time to leave. Calling for our bill, and standing ready to depart, we were surprised to find one of the girls from earlier coming up again to Felice, then speaking hurriedly with her, obviously in a tearful state. She was, Felice explained, being continually pestered by a local man; and, to make matters worse, someone on the staff was putting pressure on her to 'go along' with him. Could we perhaps smuggle her out with us as we left?

Ingenio went to call for some taxis. The rest of us formed up in a double file on either side of the girl, and walked her straight out through the main doors. No-one seemed to object, nor tried to prevent her from leaving, and in due course we waved her goodbye outside what she said was her house. Whether we were genuinely saving a damsel in distress, or whether she was merely using our gallantry to relieve her from a particularly undesirable male, or even just to get a free ride home, we'll never know. But at least it made us feel good for a while.

Soon enough, we had to bid our reluctant farewells to Felice and Ingenio. Having concluded our site studies, we returned to Bogota and presented a verbal preliminary report to the Caffeteros, concluding that further work should be undertaken by an appropriately-qualified site manager with a view firstly to encouraging farmers to begin growing cane throughout the Cucuta valley, and then, if that proved successful, proceeding with the factory's construction.

Our recommendations were gladly accepted, and T&L was asked to provide someone for the post. We sent out Saul Burnett, yet another former Trinidad 'supremo'. Both he and his wife, Joy, spoke good Spanish, so they quickly and easily integrated into the area, no doubt helped by Saul's large, friendly face very closely resembling that of the widely-recognised Colonel Sanders, of Kentucky Fried Chicken fame; hence the local nickname which he very quickly acquired.

I returned to Cucuta a year later to check on progress, and stayed at the same downtown hotel where Joy and 'Colonel Solly' had made their long-term home. By now, Saul had ventured out into much more remote areas than those covered by our initial investigations, and he was keen that I should see them. One morning, therefore, we set out along a very deserted dirt road, which became more and more broken up the further that we drove away from town. After many miles, with the road now virtually non-existent, I happened to glance at the vehicle's fuel gauge. To my momentary alarm, it read almost empty. Presumably it was just broken, awaiting spares and repair. Or so I desperately reasoned, swiftly dismissing visions of the bleached bones of forever lost travellers; or, at the least, of a very long, hot, tiring, thirsty, and, the way that my luck generally ran, far from simple, obstacle-strewn walk back to eventual civilisation.

When I offhandedly mentioned it to Saul later that evening, however, I was met with a short, pensive pause. We'd started out rather earlier than he was used to, he explained quietly. He'd forgotten to fill up the night before.

"But don't worry," he reassured me after the event. "I always carry a couple of bottles of good whisky for emergencies. If things had got really dire," - they'd certainly soon looked like qualifying as such to me - "I was fully prepared to sacrifice them."

I hadn't previously been aware that Land Rovers would run on that particular fuel, grade of quality notwithstanding: but Saul assured me that he had indeed resorted to it before. Since, however, we had actually managed to make it back to Cucuta with the needle still happily banging up against the 'E' mark, I was never able to verify his claim; or, probably fortunately, otherwise.

It was ten years before I got to pay another visit. By then, Colombia had indeed sadly achieved its reputation of being the world's drug supplier, and the cartels and barons ruled it accordingly.

My planned trip was a short one, its purpose simply for me and an agriculturalist from T&L's Miami office to review the ongoing workings of the 'new' Risaralda estate and factory, now in full production. We were met and then accompanied around by one of the operating company's directors. He seemed just as friendly as my previous contacts in the area, and saw just as well to all of our travel and accommodation needs. Yet both the American and I sensed some unease in his manner, a reservation which we couldn't quite put our fingers on.

It wasn't until it came time for us to leave that he finally drew me aside. Would I, he enquired, be prepared to accept our company's pre-agreed consultancy fee right there and then? In cash?

I wasn't entirely happy about this unexpected, not to say unusual development: but it soon transpired that I wasn't exactly being presented with many other options, either. There was currently, the director explained, some "difficulty" in exporting money from the country by way of the usual international settlement methods. Physical US Dollars, on the other hand, still worked a treat. So I reluctantly agreed to accept three thousand of them, in a selection of nice used notes, safely sealed within a large brown envelope.

My colleague wouldn't have any part of it, claiming that it was utterly prohibited for US citizens to carry such large sums into their own country: which was quite possibly true. But, he added helpfully, as I was British, I should be able to carry the money through US customs "just fine".

Should...

Luckily, that did prove to be the case: or, at least, I wasn't stopped, searched, and subsequently asked to explain my precious burden, an experience which, I suspect, could have been very interesting indeed.

Once safely back in our Florida office, of course, they were more than willing to let me to settle his share from my fat wad of bills - at the least, I ought to have deducted my own percentage fee for services rendered - before carrying the remainder back to Bromley for addition to the general T&L UK coffers.

I wasn't particularly popular with our accounts department, that's for sure. They weren't used to clients of any nationality settling their bills in such a form, and rather suspected that there might be some kind of "funny business" involved. Still, I pointed out, I had ensured that we'd actually been paid, and promptly, which might otherwise never have been the case.

They didn't stay upset for long.

****

Mexico also produced a lot of sugar. That said, with the exception of a few quick evening trips over its southern Belizean border, I only once paid it a proper visit. This was at the invitation of a cane estate situated more or less at its centre. They'd been experiencing, so they said, "great troubles" with both the loading and onwards transportation of their cut cane. For some reason, however, further precise details of their problems hadn't ever been obtained.

And thus, armed with their usual sense of adventure, if not exactly a lot of information, the intrepid, and most definitely intrigued team of Edwin and Sterling - Jerry by then our firm's senior agronomist - set forth to investigate.

Landing at Mexico City, we were immediately taken on a lengthy car journey south, descending from the high lands near the capital to the much lower areas of the Gulf state of Veracruz. The road wound down an escarpment in a series of sharp, steep bends. Our driver, with the light failing about as quickly as we were travelling, seemed quite oblivious to the long, precipitous drops on either side. We were, to say the least, very relieved when we finally pulled up, in the pitch dark, at our destination.

This test of nerve aside - and the ever-looming anticipation of our reverse trip back up to the capital - it was, overall, a pleasant and enjoyable stay. We'd been placed in a magnificent guest house, its public rooms walled with glass and enclosing a small courtyard where, right in the centre, there grew a beautiful tree, its boughs entwined by the stems and large yellow blooms of an exquisite 'Easter' orchid.

But these stunning surroundings, a perfect accompaniment to our first morning's breakfast, only belied the astonishing conditions which we confronted once driven through the fields. Never and nowhere before, in either of our extensive combined years of experience, had Jerry or I come across ground like this. True, it was only gently-undulating: but the entire surface was also covered in a thick layer of huge round stones, each as large as a football.

We did at least know what they were: basalt 'bombs', spheres of very hard rock formed when originally spat, as molten globules, straight out of a volcano's heart. With Mexico being, in parts, a volcanic country - some of its cones still remaining active - we'd pretty much been expecting to find a few related deposits, or forms of debris cover: but nothing like the size or intensity which we were encountering here. The soil beneath would, we knew, be very fertile: but just how the local farmers had ever managed to plant their cane in it in the first place was wholly beyond our comprehension.

The estate's problems, on the other hand, were extremely easy to understand. They were using mechanical grabs to load the cane, and these inevitably picked up a good number of the basalt balls, consequently carrying them to, and then releasing them upon, the factory's feeder tables. The only removal solution which they'd so far devised was to have men physically swing down on jury-rigged ropes, each gathering up and clutching one large, rounded, and very heavy lump, and then somehow, before being carried away with the cane on the moving conveyor, scrambling one- or even no-handed back up his rope and over the edge of the surrounding metal plating. It was obviously extremely dangerous work, and, more by sheer luck than anything else, many had only very narrowly escaped serious injury or death. Even with that aside, all kinds of damage was regularly occurring to the processing plant, the speed of cane throughput was being significantly compromised, and, fundamentally, the cost of continuing such an operation was proving simply untenable.

Jerry and I were able to suggest some solutions to eliminate a good number of the rocks being gathered during loading, and for refining the sorting of those still being carried into the mill with the cane. We also made some recommendations for easing their cultivation problems, and I designed a modification to their rail system which I thought might help. But whether any of these were ever fully implemented, I don't know. As with their original, rather reticent communications, we never heard back from them again.

****

It's difficult to find a more central part of the Americas than Panama, the narrow strip of land which barely separates the Atlantic and Pacific oceans; and, indeed, allows them to mingle via its famous, continent-dividing canal.

Apart from some marine traffic, however, there was little other major activity to see. There were small farms a-plenty, certainly, some growing crops, and others with cattle: but there were no large-scale or commercial crop schemes, plantations, or estates at all.

To this end, Brian McCloud and I had been sent in to ascertain what possibilities existed, from an agricultural point of view, to develop a national cane sugar industry. No overt problems of markets, shipping, construction, or infrastructure seemed to stand in the way; nor were there any restrictive government policies, or geopolitical or economic or labour troubles; nor anything else. So it was, it appeared, simply and wholly down to the two of us to determine what could and couldn't be done.

A meeting was arranged with a government official, who told us that the most likely area for such a project, considering not only the right climatic and topographical aspects, but also the availability of land, was up near Victoria, in the very north-western Bocas province, near the Costa Rican border. This was some two hundred plus miles, and a full day's drive, from Panama City. We were assigned a car and driver, and, allowing time first to study the area's meteorological and other records, arranged to travel up the following day. This also gave us a little time to watch a ship - albeit sadly only a small one, not like the ocean-going liners or huge super-carriers which sometimes came through - negotiating one of the canal's great locks.

Our destination lay to the eastern Atlantic, or Caribbean, side of the country, and the lands through which we passed on our long journey were, in the main, rather flat: we certainly didn't see any of the mountainous parts through which the canal's constructors had battled during their way towards the capital.

Having settled ourselves into a hotel, we set about our usual routine of driving around potential locations. The general topography certainly seemed suitable, and the terrain wasn't overgrown with trees. The initial plot's soils, however, didn't look promising, so we decided to investigate some land to the other side of a rather wide river: an area to which, annoyingly, if not entirely uncommonly, no current road access existed. But after some asking around, one of the local farmers gave us directions to a footbridge, by which, he assured us, the river could be crossed.

That, surely, should have rung a few alarm bells...

We knew that it would be a good day's slog just to get a basic look at what was there. Very early one morning, therefore, we had our driver take the car as near to the purported, un-mapped structure as possible, asked him very kindly to await our undoubtedly-late return, and duly strode off into the unknown.

We arrived at said crossing point just a few minutes later.

I really, really wish that we hadn't.

The considerable distance between the river's banks - together with the inevitable intervening deep gorge, of course - was indeed spanned by a what could just about, in its loosest sense, be termed a bridge: a very narrow one, suspended not by ropes, nor cables, nor vines and creepers, but just a few lengths of twisted-together wire.

Oh joy of great and abiding joys!

Still, I reassured myself hopefully, at least here someone had made an attempt at using proper materials. And everything was, at either end, properly secured - I hoped - to a solid, if basic, timber frame. And not a lot else, for the space between its slender walkway - yet again, only just wide enough for one foot to pass the other - and this all-too-thin-looking support system, already running much higher above the decking at either end than it did in its centre, was filled with nothing except the even-thinner vertical wire attachments which, every three feet or so, directly linked it all together. The decking itself, of course, in the best possible way of these nightmarish things, consisted not of regular crossways planks, but just a number of remarkably non-thick tree branches, placed lengthways through the lower loops formed by these equally-non-reassuring vertical hangers.

Overall, therefore, it wasn't too surprising to find that the structure as a whole also, like all prime examples of its type and general locale, sagged significantly towards its middle, any crossing therefore requiring firstly a descent of its distinctly uneven slope, followed by a good climb back out.

Which we then proceeded to do.

With lessons leaned from my last encounter with a remarkably-similar, if marginally less sturdy contraption, and with all essential items now firmly secured across our bodies or within our clothing (why did we never take rucksacks, I only now ask myself?), I climbed up the wooden end frame and began to cross. My outstretched arms were fully raised, grasping the suspension wires above me: but, as I descended the rickety, wriggling curve, my hands gradually fell until, at the halfway point, they were well below waist level. From there on, it was just a constant physical pull to get out of the dip and up the other side. But my arms were now again extending ever higher, soon overhead, and the further that I moved up the slope, the more the effort became to haul myself up.

The more, also, that a rather major design flaw became apparent. The whole suspension system was made not just from wire, but from barbed wire: two or three strands of it, in fact, all woven together on either side, with a consequential proportionate increase in the number of protruding barbs per any given length. Instead of being able simply to slide one's hands along, therefore, and thus maintain a constant, reassuring contact in the face not only of the steep and generally-insecure incline, but the rough, uneven branches underfoot, ever-ready to trip up the wary traverser and tip him straight through its non-existent sides into the yawning depths below, it was necessary to let go and retake a fresh, firm hold at every single step, ignoring the inevitable, extremely-painful, but wholly-essential puncturing of one's fingers and palms in the process. Still, at least the effects of vertigo due to the extremely-excessive amount of open spaces around this swaying sometime-structure, never mind the sight of the dark, frothy waters rushing past far below, occasionally gave the mind something else to focus on; and, as necessary, vice versa.

We also both tried very hard not to dwell at this point - on every literally-bloody barbed point - on the far-from happy realisation that, no matter where our wanderings might take us during the day, whatever prospects lay ahead for us - blood poisoning aside - whatever delights we might discover, whatever mere satisfaction we might take from our knowledge of the conscientious lengths to which we would go in fulfilling the requirements of our company's contract, we still had to come back the same way.

Memo: it doesn't matter how hot and sticky it is, always pack a good thick pair of leather gloves.

Rather than endure another tedious, time-wasting drive back to Panama City, we rearranged our return journey and flew back down from the local airport.

That wasn't any better an idea, either, as it turns out, our final day's near-disastrous flight almost, as far as we saw it, proving to be both our final flight and day, full stop.

Both Brian and I swore never to fly Panamanian airlines again. Which wasn't a particular problem, as it turns out, because neither of us ever went back.

12

BRASILIAN BACCHANALIA

Brasil (or Brazil, if you prefer) is a huge country, its topography and climate varying from the almost impenetrable north-western rain forests to its cold, southern districts, via virtual deserts, bleached white sand dunes, and the palm-fringed beaches of its long, eastern coast.

"O linda!" - how beautiful! - exclaimed yet another boatload of Portuguese adventurers, arriving for the first time upon its forested north-eastern shores; or so the story goes. The very old, and very pretty town of Olinda (officially founded 1535) still sits upon one of the promontories where these explorers apparently first landed in 1496 (or thereabouts: but certainly not, according to the locals, the official 1500). The Portuguese words for the bronzed colours of the native, orange-red hardwood trees which were then so plentiful - the 'brassa' of the 'pau-brasil' trees - also supposedly seem to have led to these newly-colonised lands becoming known not by their official title of the 'Land of the Holy Cross', but as the more colloquial 'Terra do Brasil': probably a good thing, since there's far too many Santa Cruzes around already, and it would doubtless all have got terribly confusing.

Historic nomenclature aside, a good deal of said beauty, particularly as regards the country's rivers, was, by early 1971, suffering severely under the mounting pressures of industrial effluent pollution, stemming mostly from a multitude of rum and other distilleries. It was already illegal, at least in some of the country's individual state regions, to dump their waste products directly into the water courses: but fines for such transgressions were minimal, if even in reality imposed, and formal regulation, in practice, was simply non-existent. The intention of my first - and most definitely not so run-of-the-sugar-mill - visit, therefore, was to persuade the Brasilian government to try out a system which we'd developed in order to deal with the issue.

My first test, after undertaking several surveys of various offending factories throughout the north-eastern state of Pernambuco, was to accompany Jean Ryker, our Rio de Janeiro agent, to explain the proposed system to the managers of a number of other plants. Generally, they all seemed to welcome the idea. It was then time, courtesy of several introductions already gained through Jean and other T&L staff's previous meetings, to try and meet as many other assorted useful and influential Brasilians as possible; including, soon enough, the minister of industry himself.

Senhor Pratini de Moraes listened very carefully to what I had to say, admitted that there was indeed a considerable problem, and then declared: "But I like effluent! If there is effluent, then it means that we have industry. And if there is industry, it means that we have employment. We have a big country, and there is plenty of room for the effluent to go. There is not so large an area in which to employ people."

He had, at least, a point.

Unfortunately, as far as our scheme went, it was also a clear sign that any further progress would be highly unlikely, since none of the individual factory owners was prepared to expend the necessary effort, not to mention the considerable sums of money, required to deal with the issue without there first being some form of compulsion, if not actual added incentive, forthcoming from the government, too.

Undaunted, I was back again within a few months. I was still based in the waterfront city of Recife, at Brasil's most easterly, and most obviously-prominent point; and I still had quite a novel brief. This time, I was to investigate the possible formation of a joint, three-party enterprise between the north-eastern arm of a São Paulo based steel-fabrication group, a local iron foundry, and, of course, Tate & Lyle.

It became, as Brasilian business matters so typically turned out, a rather longer trip that I'd planned: some tiring three months spent mostly sitting and talking, initially with the other two companies' management teams, and then with representatives of an accounting partnership of British origin, Price Waterhouse Peat do Brasil, a firm which, amongst many things, retained three legal partners with the sole, all-consuming task of keeping everyone fully up to date with the country's constantly-changing laws.

This time, however, the outcome was fruitful. Our new enterprise aimed to combine T&L's consultancy expertise with the specialist manufacturing capacity of its other two partners, and thus provide advisory and design services for the overall sugar industry throughout the entire north east of Brasil. We christened it 'Talycopi' - an inventive combination of Tate & Lyle, Cosinor and Piratininga do Nordeste - and it was to be based at Cosinor's central Recife factory. Yours truly was to act as its resident T&L representative, and also its overall manager.

My stay was initially planned as a six month posting, but, due to various start-up difficulties, very soon extended to twelve. This, combined with the city-based location, was a long enough time to warrant bringing my wife and seven-year-old daughter out to join me. It wasn't an undertaking without its concerns, though: Eleanor, whilst admirably used to coping alone during my many absences, had never even travelled abroad before, and Rai suffered very severely from several health problems which necessitated frequent urgent medical attention, ongoing consultations, and sometimes hospital stays. But we decided to take the plunge, and so, in June 1971, with our house and its contents soon safely - so we hoped - to be placed in the temporary care of our neighbour's son and his friends, I went back to Recife along with Andrew King, my chief executive director, to finalise the details and begin my work.

Cosinor had appointed their own counterpart local manager, one Luca Afonso Vitor Ferreira Pinheiro da Silva, to give him his full, Brasilian-style name. At times utterly serious, at others ebullient and charming, Luca was, as I soon found out, both a very unusual and a very useful man, well-known and widely-connected not only within our target areas, but throughout national government and various other country-wide institutions, too. Armed only with a small card and some kind of official badge, he also possessed - and soon repeatedly demonstrated - a strange, occasionally almost supernatural power to gain us entry into any place which we wished or needed to visit, and also to obtain prompt, if not usually immediate, attention and service from any government office or travel company which we might for any purpose, and at any time - Brasilian definitions of such notwithstanding - require.

One night, for instance, after a long and busy day down in Rio, Andrew, Luca, and I headed back to the airport to catch the last flight back to Recife. Given its size, Brasil had already developed such a fast, regular, and surprisingly-reliable network of internal air connections that one could quite easily use any of the four or more local and international carriers to travel pretty much at will between its various major cities and business hubs, flying somewhere simply for a morning or afternoon meeting, and then returning home the same day.

Arriving at the ticket desk this time, however, we were told that the flight was, most unusually, already full.

"Não há problema," came our escort's confident response. It was a phase which I was to hear from him time and time again; sometimes, several times a day. "Just wait here."

He then disappeared off somewhere behind the airline's official backdrop, only to reappear shortly afterwards, beckoning us to follow. Standing in a small private office, we were promptly provided with our boarding passes, and immediately invited to join the flight. Someone else, presumably, didn't make it back home in time for dinner: but we, at least - feeling mildly guilty, it has to be said - did.

It was only some time later that we finally learned just how important a man Luca actually was in contemporary Brasilian circles, being a graduate of, and indeed already a past president of the very select association of graduates of, the 'Escola Superior de Guerra'. Sometime in the late 1940s, the country's 'newly democratic' government had established this organisation to function along the lines of a specialist academic institute of advanced postgraduate studies. Its aim, despite the name, was to train certain 'elite' individuals, if and when called upon, to be able to assist in running the nation, taking on strategic management, leadership, and advisory roles within the areas, mainly, of development, security, and defence. Although stemming from military roots, only a third of those selected now came from there, another third being drawn from the political classes, and the remainder from various educational, professional, and technical personnel. Luca had qualified by being both a university Professor of Agriculture, and also by having previously had a military career.

I began my residence in Recife at the small, single-storey Cote D'Azur hotel, set beside the main beachfront road about two miles away from the main river bridge which led into the city's brightly-lit centre, complete with its sprawling office blocks and ring of steaming factories. My little home-from-home had no window, being fully enclosed by other rooms, no TV, and little else to recommend it apart from a soothingly-humming air conditioning unit. I therefore didn't usually bother returning until fairly late evening, when I might lie and read for a short while before turning off the large main overhead light by means of a two-way switch set next to my bed.

I hadn't actually spent that many nights there when, at about one o' clock in the morning, I was woken suddenly by a bright flash, a loud bang, and the sounds of a dog's pitiful howling. There followed a long period of complete silence - the air conditioner had obviously stopped working - and total darkness, as a trial of the light switch soon proved. A power failure, I quickly deduced. Nothing, one way and place or another, entirely unfamiliar.

Soon, however, I began to hear anxious voices from beyond my room, quickly followed by more pressing and urgent enquiries, accompanied at first by a few tentative knocks, then much more forceful banging on my bedroom's door. Reluctantly, I sat myself upright, reached out for the small battery torch which I always kept on my bedside table, and switched it on.

What I saw wasn't exactly what I'd been expecting. Both the bed and surrounding floor were covered by fragments of broken glass and once-white plastic, whilst, beside the door, half way up the wall, there was now a gaping, blackened hole, with still-smoking strands of burnt electric cable hanging out.

Having then very quickly, if extremely carefully, got out of my bed, and after assuring the hotel staff in the corridor outside that I was uninjured - a fact which, given the shattered glass alone, actually seemed quite miraculous - I was politely requested to vacate my room. It was, apparently, now considered unsafe. I wasn't, in the circumstances, about to argue. I quickly packed up my possessions, and relocated a few doors down.

Everything was explained the next morning. Engineers had been carrying out some late-night work on a pavement transformer box just around the corner, and had temporarily left it unattended. The wide-open doors had quickly proved far too tempting for a local stray dog, who'd wandered straight into its innards and, presumably, started to do what dogs seem to like to do in such situations. The resultant cross-linking of the incoming high voltage and usual outgoing 220v supplies caused an immediate surge into the local distribution network, and, for reasons of Brasilian electrical supply design which I shan't bother to elaborate further, my room just happened to contain the nearest attractive circuit in the vicinity for those wandering 11,000 or so Volts. Very fortunately for me, it was the ceiling light fitting, the switch next to the door, and the cables feeding the two of them which had taken the full force of the blast, and not my bedside controls.

The dog, not surprisingly, died.

This wasn't, by far, my last experience of Brasil's peculiar electrical problems. Not all that many days later, in fact, I accompanied Luca on a visit to the city-centre offices of an important and influential group of sugar factory owners. It was an unusually windy day, and, as we approached the large tower block, the heavens also decided to open, pouring down their best impression of a full-strength tropical storm. The electricity in that part of Recife was still supplied by means of ancient overhead cables, many nothing but bare copper strands with no insulation at all. The heavy rain and strong gusts, of course, immediately caused the wires to touch, shorting out all of the area's power.

"Não energia," Luca shrugged, as we stepped into the darkened foyer, complete with inoperative lifts. Followed, of course, with the usual cheery statement: "Não há problema. We will walk up."

And so we did, all fourteen floors, complete with several bulging briefcases and only very restricted emergency lighting.

No-one seemed to find our breathless, profusely-sweaty state at all unusual when we eventually presented ourselves for our meeting. They simply handed us the standard strong, sickly-sweet coffees - water would have been much more preferable, in my opinion - and ushered us in.

Three hours, and six jolts of caffeine later, after having finally settled down to accepting that the long rounds of violent verbal dialogues, gesticulations, arm-waving, angry-looking faces, and raised interjections every time that Luca translated one of my questions were nothing more than their normal version of "hello, yes, very pleased to see you, happy to try and help", and responses of that ilk, it was finally proposed that we would all convene again the following day.

Was there now some kind of problem, I asked hesitantly?

"Oh no! Não há problema at all. Everything is actually very successful. But now that we have introduced you, we can plan to start discussing business."

I was very quickly to learn that Brasilian commercial practice wasn't quite done in the manner which I was used to, either back in the UK, nor in any of the other countries within which I'd thus far worked. No matter what the occasion, before anyone actually, as it were, started getting down to business, it was first of all customary to talk at length about the health, or otherwise, of all of the participants' relations, as well as that of any shared, or well-known, or just generally-interesting, friends. There then came another suitably-long period of apparently-random personal reminiscences, followed by an appropriate amount of simple shared observations about the ongoing state of the country, and finally the wider world as a whole. With everything else exhausted, one might just then possibly consider getting down to the real purpose of the visit: if you could still remember, or be bothered to care about it, any more.

In the meantime, either to stimulate conversation, or perhaps the underlying cause of it, or maybe simply to keep everyone awake enough to get to the end of all this palaver, a great many 'cafezinhos' would have been drunk. These very dark, very strong, and very bitter coffees were served without milk, and in a small cup which had first, prior to pouring, been half-filled with sugar. Rather like the verbal preambles, their supply and consumption throughout the entire course of any given occasion, at any location whatsoever, seemed utterly without end.

In these early days, I had little to no grasp of the Portuguese language, of course, and so, in my many preliminary meetings, never mind all of their accompanying personal verbiage, I soon found myself completely in the dark, mentally if not always literally, and utterly reliant on Luca, able to do nothing but trust that he was accurately and properly conveying both sides of our discussions.

I did, however, already have a fair command of Spanish, which technically has a lot in common: so maybe, sometimes, things weren't quite as dire as I liked to make out. It also, to my obvious benefit, ought to have worked the other way round: except - and it was a very big exception - that the Brasilians, although many surely knew the language well enough, simply refused even to listen to any Spanish at all. Rather like the French, really. Still, armed with a 'Teach Yourself Portuguese' book, and with an ear very quickly becoming forcibly accustomed through our many, nay ceaseless, associations with local businessmen and other contacts, it wasn't actually that difficult to 'convert' to a reasonable degree of understanding.

Absolutely no language or conversational skills whatsoever were required for another aspect of general Brasilian practice, though: one which I initially found not only disturbingly alien, but, for a good while, more than difficult to accept, and even harder to reciprocate. This was the customary mode by which all men greeted not only their friends, but all of their known commercial associates, enthusiastically throwing their arms around each other at every possible meeting, before entering into a sustained session of vigorous, close-quarter back-patting with whichever arm lay over their partner's shoulder. Depending on how long it took for both to voice their various additional salutations and other appropriate queries and observations, this whole highly-embarrassing embracing process could often last for several minutes; or even more. Thankfully, at least, they hadn't yet adopted that continental kissing business as well, and I eventually managed to conform sufficiently that the Brasilians didn't continue to think me quite so odd as I'd obviously first appeared. Nowadays, perhaps, this kind of behaviour wouldn't strike us as so unusual: but no-one has yet, to my knowledge, perfected its drawing out quite as well as the Brasilians did.

It quickly became apparent that time was also very different in Brasil; the appreciation of its existence and importance, at any rate, especially when concerning appointments. This laid-back attitude initially, and continually, not only irritated and frustrated my British mindset, but frequently became a significant impediment to getting work done at all. Luca would, for instance, tell me that we were to meet with someone, or perhaps go to a factory or an office, at a particular time. I would enter this into my diary and, in due course, prepare accordingly. Sometimes, we would set off in good time to keep the appointment; at others, I would still be sat in the office, awaiting my escort, long after the scheduled event had passed. When he finally made his appearance - and he always did, at some point - Luca would then get on the phone to whoever it was, give his excuses, remake the arrangements, and - "Não há problema" - off we would then, somewhat belatedly, set.

It wasn't just Luca, either: it seemed to be a Brasil-wide issue, and it finally came to a head not too long after Ellie had arrived. Luca and I were visiting a factory about three hours' drive from Recife, having booked to meet its manager at two, "or thereabouts", in the afternoon. We were intending to demonstrate a new chemical process which Tate & Lyle was introducing into many countries for the home-based production, rather than currently-common importation, of fully-refined white sugar. (You can argue the pros and cons as much as you like, but this, at the time, was mostly considered to be a majorly good thing.) The actual procedure would take less than an hour. Even allowing for some Brasilian procrastination, I was confident of returning home by early evening.

Arriving, at my insistence, at the planned time, Luca and I were shown into the manager's office, and, after the usual long-winded greetings and enquiries, we were given a very thorough tour of the factory. We would then, I'd expected, get down to at least the preliminaries of our business. But no: for then, and only then, and for no apparent reason, the manager and his chief chemist informed us that they immediately had to go off to some other appointment.

Time went on, thumbs were twiddled, a few of the remaining alternative staff members were chatted to for a while, several cafezinhos were consumed, and I began to wonder when, and indeed if, I'd ever be able to proceed with our primary purpose. I persuaded Luca that we should at least visit the laboratory and prepare the necessary equipment: which we duly did. And then we went back to the office, drank many more coffees, and waited again. We reached the time by which I'd originally anticipated being happily back home; and still, here we were.

The manager did eventually return. Not so much as the slightest apology was ever offered for leaving us for so long. In lieu, however, we were most jovially invited over to his house for dinner. It was going to be a long night! First, I knew, we'd have to endure a lengthy polite conversation with every member of the family present in the household - doubtless quite a few, given the usual large family groups - followed by several pre-dinner drinks, and then a very leisurely-taken meal.

And so it indeed was. We finally finished our food at a little before midnight, expressed our long thanks to all concerned, bade our even longer farewells to the assorted, multiplicitous relatives, and then drove back to the factory's laboratory. The chief chemist, needless to say, had long departed: but there remained a duty bench assistant, who said that he could reproduce the demonstration equipment which we'd set up so many hours ago, and which, in the meantime, had all been dismantled again and put away.

Only at that point, therefore, by now the small hours of the morning, could I proceed to explain in slow, carefully-translated detail what I was about to do, and how I would soon produce from the raw extracted syrup - normally simply boiled up straight away to obtain a base, unrefined, light brown product - a clear liquid from which "pure and clean" white sugar could then be crystallised out. This all took some considerable time, but I did eventually get around to commencing, and, after the necessary further hour or so, completing the actual display itself.

The results, to my yawning relief, were more than satisfactory. The manager seemed very impressed. And Luca and I at last began the yet-more-time-consuming process of making our final goodbyes, before setting off on our three-hour return drive back: in my case, to one severely-worried wife.

I not only immediately insisted that we had a telephone installed at our flat, but the next morning blew my proverbial top, telling Luca in no uncertain terms that, from now on, all meetings and similar arrangements, no matter where or with whom, would be conducted according to 'English time' whenever our company's English representative - me - was present. However he understood and communicated this onwards, I'll never know: but there were never any major problems or delays again.

I am, of course, somewhat getting ahead of myself here, for, whilst getting to grips with all of my new world's peculiarities and customs, not to mention simply dealing with our new company's many start-up demands, I still had to make everything ready for another item of imminent importance: the arrival of one wife and daughter.

Cosinor had given me an open hand to seek out a suitable home, and to furnish it as we pleased. All - all! - that I therefore had to do was to draw up a full list of everything which would be required, down to the very last, smallest item, and then to go and procure them in person. In that day and age, communications with home were not cheap, nor always instant or easy, so whilst Eleanor had tried her best to educate me before I came back out, there were many things which she or we had previously forgotten or overlooked, and upon which I had to make an executive, on-the-ground decision without further consultation.

Happily, not to say helpfully, Cosinor had also seconded me the assistance of one of their clerks for this particular exercise. Benico was a man not only of surprising efficiency, but also with a very good command of English. Together, we soon found a suitable apartment on the top floor of a newly-built, flat-faced, four-storey building of eight flats in total. It was set upon the Rua Antônio Falcão, a pleasant, wide, residential road which was apparently named after a man - exactly who and why I never discovered - who had lived just around the corner. More importantly, it was just one short block away from the main beach-side road. Its front windows also looked over onto the square, high-walled garden of a grand old house which accommodated some kind of mysterious 'very important Brasilian' - so secretive that, even with Luca's connections, we never actually managed to establish exactly who - the compound's heavy, metal-shuttered gates constantly guarded by several armed, steel-helmeted military police. Somehow, though, it always made us feel rather safe.

The large, marble-floored entrance hall of our more modern building was also continually guarded, this time by one of two wonderfully-complimentary, casually-uniformed 'zeladors'. No matter at what time of day or night, these resident caretakers, receptionists, handymen, and cleaners - little and large, tall and thin, short and stocky - were mostly encountered slowly shuffling their way around the spacious lobby's smooth, already-spotless floor, polishing it by means of the two thick, bunched-up rags being alternately swirled around underneath their bare brown feet. Neither Pedro nor Fernando ever spoke a word of English: but they always went out of their way to be friendly and helpful in so many ways.

Helpful, too, as it unexpectedly turned out, was the fact that the elderly and quite reclusive couple who lived on the floor below ours were, once upon a time, English. They'd been resident in Recife for many a year now, and were confirmed Brasilian citizens, but were still able to give Ellie and me some invaluable advice and pointers when we required them.

Accommodation secured, I was then free to go shopping for all of the furniture, fittings, furnishings, carpets, curtains, fans, fridge, cooker, and anything else which I've doubtless now, as then, forgotten. Directed to a huge superstore in the city centre, I was relieved to find that it sold pretty much all that I needed: so along hurried Benico and I - several times - and slowly ticked our way down my very long list. I prided myself that I'd finally thought of and obtained everything by the time that Eleanor arrived just a couple of weeks later. I didn't, at least, receive too many complaints, and the bright-red sofa and chairs were just the only ones which they had in stock and could have delivered in anywhere near the timescale required; albeit also being, I was assured, extremely fashionable for the time.

My final task was to make contact with the Escola Americana do Recife, an English-speaking, American-run grade school situated at the city's outskirts. Enrolling Rai with surprising ease, I also arranged for the school bus to collect her each morning from the roadside just outside our flat. Of course, although her first day's journey to school went smoothly and entirely to plan, whilst making her way back home, caught in the excited midst of all her new classmates, she just had to manage to get onto the wrong one of the two awaiting vehicles (no-one ever told me that there was another!), one a sleek, modern, air-conditioned coach - for all the non-Americans, it transpired - and the other an old, proper US-style school transport. It was by means of the latter, therefore, that she was taken, wholly unsupervised, all the way to the far side of Recife, and there, eventually - once her lonesome plight had finally been recognised by the non-English-speaking driver, and the appropriate messages somehow relayed back - left to await collection by her extremely-worried parents.

Annoyingly, pending a few final arrangements, I was still resident in the cramped Cote D'Azur on the day that Ellie and Rai were due to travel out. I was also already pre-booked to go to Rio on a business trip. I therefore thought - for some stupid, but doubtless well-meant reason which wholly escapes me now - that instead of either rearranging some part of those plans, or at least having my family met at the airport on my behalf, it would be a nice idea if they initially joined me down there, and we all then travelled back to Recife together. Accordingly, I had their airline tickets made out all the way through to Rio, and instructed Ellie to travel on to there.

Unfortunately, what I forgot to emphasise sufficiently was the small matter of the words 'through' and 'on', relating to the fact that the outbound plane from London called first at Recife, before proceeding onwards to Rio, and then São Paulo. So inevitably, come the appointed day, there I was, still waiting patiently at Rio de Janeiro airport whilst, as eventually ascertained after some fraught calls with Cosinor, my wife and daughter were already installed in their new, temporary home. I immediately caught the next flight back north, and, although the start to their new life hadn't exactly been the best welcome that I had hoped for, we were soon reunited.

I never fully understood what had happened, but someone, presumably back at Heathrow, noticing that the ticket was ultimately made out for the eventual additional return leg, had somehow ignored the initial 'via Rio' portion, and handed out boarding cards tagged direct for Recife, also ticketing their baggage accordingly. Eleanor, tired and anxious after her first experience not only of long-haul air travel, but of any form of international or air travel at all, was therefore bundled off the plane at 'Guararapes International', complete with child and heavy baggage, but without any further assistance.

Matters then weren't made any easier by the fact that Rai, for medical reasons, couldn't be vaccinated against smallpox, a precaution still mandatory due to an ongoing outbreak somewhere else in the world. But we'd already prepared for this, with Eleanor carrying an explanatory letter from senior doctors at London's Great Ormond Street children's hospital. What we stupidly hadn't thought to prepare for, of course, was the small fact that, when she submitted the letter at Recife airport's immigration control, no-one could read it in its English form. Someone eventually found an airline stewardess who had enough language skills to translate, and at last that problem was overcome.

But then, Ellie realised, in the absence of one expected husband, she didn't know where to go next. Moreover, she had no Brasilian money to pay to be taken there: in those days, everyday people didn't have charge or credit or debit cards, there weren't such things as ATM cash machines, and I simply hadn't thought to tell her to get any travellers' cheques because, of course, I would be there to meet her. Once again, fortunately, the helpful stewardess arranged the exchange of some Sterling notes into Cruzeiros, and then called a taxi to take the exhausted pair to the Hotel Cote D'Azur, last known correspondence address and assumed current resting place of one Mr - "Sen-yor?" - Edwin.

Yet upon their arrival and subsequent enquiries at the reception desk, the hotel clerk had only smiled sadly back, and indicated that he knew of no such person. Utterly dejected, without anything else to do, and with no immediate means of contacting me that she could think of, Ellie simply sat in the reception area until the clerk suddenly looked up thoughtfully, and finally exclaimed: "Ah! Doutor Franco!" Now I'm not, and have never been a doctor of any kind, 'proper' or otherwise (take that argument up with my two PhD grandchildren), but, in accordance with yet another of those strange Brasilian practices, it was an honorific seemingly applied to anyone whose occupation was of an appropriately-respected professional or technical nature. And thus, albeit also only ever being applied to one's first name, it was how I was always introduced and greeted, from the most senior business associate to the 'mere' hotel clerk; and, fortunately, the latter was clever enough to put two and two together, and realise that the English businessman whom he knew only by that name must somehow be associated with this English-speaking Senhora.

Ellie and Rai were then shown to my room, and settled in, via a few more frantic phone calls and messengers, to await my eventual return.

We stayed only a few more days in the hotel before, with the last of the furniture delivered, formally moving into our flat. From then on, our Brasilian life began in earnest.

Rai, her health much improved by the climate, went off to school every day, although she quickly announced herself to be very bored, and initially refused to go back after the first term. The American system, doubtless combined with the twenty seven other nationalities of pupils which the EAR then taught - Rai was their sole British scholar, throughout all the age groups - seemed to be quite behind the UK's when it came to expected attainment standards, so we had to arrange for her to attend some lessons with the older children in a higher grade. Still, she soon made several friends who lived reasonably nearby, particularly a cheeky, curly-haired American lad whose family lived in a big white house a little way down the beach-front road, and a lovely young Danish girl who, being the daughter of a frequently-relocating diplomat of some sort, already spoke seven languages fluently at that very same age. She really put us to shame!

We, in turn, soon met many of the parents: a very mixed bunch, all temporarily or permanently located in Brasil for varying work-related reasons. Through them, we not only attended many social functions, but regularly joined in alongside other English-speaking local residents - mostly of US or Canadian origin, but also including some Germans - at the ecumenical Sunday morning church services which were held at the EAR school. The word 'ecumenical' here had real significance, its all-encompassed denominations ranging from the Roman Catholic, high and low Anglican, and various forms of the 'non-conformist' Baptist and Methodist churches, through to very some obscure sects and faith groups indeed. Yet they all somehow managed to come together under the general banner of Christianity, simply because there was no other place for them to worship regularly in their native, or at least second, language. They had to organise everything for themselves, too, including who should conduct the services, and who should preach. This, however, worked out wonderfully, and there was never any discernible discord between the many differing strands and shades of belief. Perhaps a lot of things would have been better if this particular Brasilian model had been able to be exported worldwide. (There was, incidentally, also a dedicated Anglican church in Recife, and its vicar usually tried to hold one English-speaking service at some point every month. Whilst not C of E ourselves, we nevertheless occasionally also attended there, and were made very welcome, just, if nothing else, for a change of scene.)

For lighter and more general entertainment, both the locals and expats were, we soon found, great club-goers. Social clubs, that is, each founded around a particular sport, activity, or other common bond, such as the Club Athletico or Club Nautico. Our firm favourite weekend haven, however, was the British Club. The atmosphere was indeed very British, despite there never having been any real colonial influences in Brasil. The cricket field was beautifully watered and mown - if sadly never used for anything but football matches - and on the terrace around its swimming pool one could always obtain toast, afternoon tea, and cakes. Its membership, however, did not live up to the name: we only ever met two other English families there. Then again, we were somewhat surprised to discover, they did constitute an entire two thirds of the known total of British families then resident in the entire Recife district; and we, of course, made up the other. One of these men, we soon found out, also carried out additional part-time, voluntary duties as our UK consul. His main official business was taken up with British sailors whose ships called in at the port: but it nevertheless provided us with some kind of comforting formal presence, just in case.

And we clearly did need some support and safety in our paucious numbers. Not long before our final return to England, we received an invitation from the club to join them in celebrating the (our) Queen's birthday. It was a lovely event, certainly, and very well attended; although, yet again, the sizeable numbers crowding the club's pool and grounds consisted entirely of Americans, Canadians, Very Important Brasilians, and a sum total of half a dozen Brits, children included.

Our introduction to the British Club actually came via my secretary, Cosinor having recruited a young English-speaking lady by the name of Helga Gianelli. She, to add to the international mix, was actually of Anglo-Austrian-Indian extraction; her husband, Tomass, was an Italian-Indian, sent out from Madras to represent the (originally-Czech) Bata shoe group. They had a daughter about the same age as Rai, although she attended a different school, and so, between work and play, we soon all became good friends, and often socialised together.

Ellie and I also received frequent invitations to dinners or parties being held at houses belonging to Cosinor's owners, their family members, and many other Brasilian business and social contacts; and, of course, Luca. These were quite standard and normal affairs for the day; except for one thing. Despite Brasil, and indeed Recife, possessing many distilleries, brewers, and bottling-plants for every imaginable type and brand of refreshing or intoxicating beverage, we were amazed to find that every 'high-class' Brasilian invariably, and only, drank whisky when entertaining; and, of course, only the most expensive brand then available, Chivas Regal. It was presumably nothing but a statement of social status, as, to my pallet, the locally-made King George brand seemed pretty good in itself. This judgement was, in fact, in due course tested, and apparently proved good, when the chairman of our T&L company came out to Brasil, and paid a passing, unplanned visit to our flat. He was supposed to be somewhat of a whisky connoisseur, but I only had a bottle of KG in the house, so I had to serve him that. After holding his glass up to the light, swilling its contents around, sniffing, and finally sipping appreciatively, he proceeded to complement me on keeping in such an excellent stock. If only the locals knew!

Personally, however, when given the opportunity, I opted for a local drink - perhaps cocktail might be more appropriate - known as a Batida. This was made up from a mixture of sugar, lemon, or sometimes other forms of fruit or coconut juices, and cane brandy, the latter locally-produced and bottled at various small farms. Served ice cold, it was very refreshing indeed. It was also near lethal. I've never been a heavy drinker, so it didn't take much before the troubles of the day were completely forgotten: but sometimes, especially after a day of meetings and all of the trappings and timing issues which went with them, not to mention coffees, becoming 'Batida-ed' was a very welcome relief indeed.

Back on the domestic scene, things generally seemed to be going well. My working arrangements meant that I could usually return home at some point every night, although there were still a few visits, both within and without Brasil, which necessitated me leaving home for days at a time. And Ellie seemed to adapt quickly enough to what was, in many ways, a very strange and different life. She wasn't exactly blessed with a great gift for languages, to say the least, and she didn't know how to drive. There was, therefore, very little for her to do all day except to attend to whatever domestic chores and shopping she chose, to enjoy walks on the beach and around the local area, and occasionally to gain some female companionship (aka meet the other wives) from within our various social networks.

She was, for a while, quite regularly entertained by a group of American ladies from the EAR church. They once took her on a day trip some hundred or more miles north, to the town of Natal - complete with its Los Tres Magos (three wise men) hotel, and many other connections with the birth of Christ - to visit the Hope, a travelling charitable hospital ship which was temporarily moored there, providing free medical services to the area's poor. But whenever and wherever she met them, these women, so I'm told, were always most openly intrigued by Ellie's 'accent'. One of them taught English at Recife's university, and, one day, along with a group of her students, they all happened to meet on the local beach. Seizing the opportunity, the American asked her British counterpart again and again to say something in her native tongue, so that her students could hear how this strange "original English" (now only mildly Midlands-tinged) sounded. There was great hilarity all round, apparently; except in Eleanor's case.

She didn't spend much time with those ladies after that.

We had quite a few other visitors round to our flat over the months, too; mostly T&L employees or contractors, mostly in the evenings, and mostly in search of a quick rest, meal, and freshen-up whilst passing though; or just a few simple hours of social company, and a good drink or few; or, occasionally, emergency settee accommodation when they couldn't secure a hotel. One man, for whom we almost felt quite sorry, was on his way back from Rio, where, falling asleep after a swim on the famous Copacabana beach, he'd been robbed of all his money and travellers cheques. He actually ended up staying with us for several days whilst the various phone calls and arrangements were made; and, we hoped, learnt his lesson about the appropriate security and back-up measures which, sadly, the foreign visitor always had to employ.

Shopping, generally, was nowhere near as difficult for Eleanor as I'd feared. For a start, there was a sizeable local supermarket, Comprabem, conveniently situated just around the corner from our new home. (In fact, our back windows looked out over its rear yard, so we could see all the skips into which unsold meats, fruits, and vegetables were disposed; and also, sadly, the many desperate people who, under cover of darkness, went scavenging there for food.) It was bigger, brighter, and much more widely-stocked with all sorts of useful household and everyday items, edible and otherwise, than anything which we'd yet seen back in south London. And it was fully self-service: something else which we hadn't yet become used to. There was therefore no need to speak Portuguese in order to buy whatever she wanted, dress-making fabrics and the like included, and the friendly staff all quickly came to recognise the 'point, smile, and pay' lady, and did what they could to assist.

We also soon had help from various English-speaking people whom we'd met, including a woman who owned a 'fazenda' or 'granja' - akin to a small-holding or allotment - and a farm shop at which she sold all types of poultry products. She also spoke fluent Brasilian, so, early on in our stay, took Ellie along to a local open-fronted bakery, where she proceeded to tell the proprietor about everything which the 'new lady' might in future need. The shopkeeper not only had a good memory, but was obviously a keen believer in good customer service, for, at each visit thereafter, all that Eleanor had to do was to indicate, and then nod or shake her head at the appropriate juncture, and whatever she wanted was, with full and immediate priority, wrapped up for her, no matter however many others - mostly the maids from other households - were waiting in line.

There were a couple of other very good, and again, to us, quite novel systems in use at the Comprabem store. The first required you to hand in all of your bags to a counter set just outside the store's entrance, whereupon you received a chit for reclaiming them after the check-out. This was to prevent shop lifting, and it apparently worked a treat. The second was a generalised glass bottle deposit scheme (plastic wasn't then in common use), applying to the large selection of many branded and local beers and carbonated soft drinks on the shelves. The general concept of exchange and refundable deposits wasn't new to us in relation to the glass milk or 'pop' (fizzy drink) bottles which we usually had delivered in England, and sometimes also bought direct from our smaller shops: but we'd never come across such a type or scale as this. Next to the outside bag depository, there was another dedicated stall, where one firstly returned one's 'empties' and received the appropriate number of 'fichas' (tokens) in exchange. Once at the till, the fichas could then be redeemed on a one for one basis against the purchase of full bottles. (Interestingly, in a more refined, automated, and hard currency or paper chit form, this kind of system now seems to be in widespread use in several European countries; but still not, at the time of writing, in the UK.) But these fichas weren't only confined to Comprabem, though: they were of a common design in use amongst all drinks retailers, even street vans and vendors - and in Brasil, a fondness for ice-cold bottles of Coca Cola, Fanta and a few other local 'sodas' meant that they were pretty ubiquitous - one's pockets often being found full of the large, heavy-duty plastic, reddish-purple discs, best described as resembling a hollowed-out UK national speed limit sign. In some cases, particularly between youngsters, the ficha had in itself become a form of readily-exchangeable currency for all kinds of other tempting goods.

Of course, come our very last visit before leaving the country for good, there was just one small difficulty. Handing over the empty bottles as usual, I, in my best penny-pinching way, requested a cash refund. The supermarket had apparently never been asked for this before. Management had to be called to resolve the matter. I did get my money back.

This wasn't the only time at Comprabem when things didn't go quite as smoothly as they otherwise should. But then again, not everything in Brasil always went so smoothly, either, and you couldn't blame those who were just trying to cope as best they could. The Brasilian currency, as a major case in point, had been devalued a short time before my first trip out. New banknotes had been issued, the Novo Cruzeiro confusingly replacing the old one of the exact same name, only worth a thousand times less. For some reason, however, the old notes weren't immediately withdrawn from circulation as legal tender, and both types remained in common use throughout a lengthy, and thus quite confusing transition period. They had, at least, managed to withdraw the old coinage: but they hadn't yet managed to mint enough of the equivalent new 'Cent' (Centavos) coins, particularly in the lower values.

The whole thing proved rather problematic when Ellie first went out shopping, since she was often asked for payment in amounts of 'Cruzeiro Mille', as the new version became known, for goods only worth - and, indeed, marked up on the shelves as costing - a thousandth of the demanded price. Until the proverbial penny dropped, she thought that either a whole lot of things had suddenly become extortionately expensive, or that, much more likely, she was being taken extreme advantage of. The latter suspicion wasn't helped in the slightest when the till assistants kept pushing unwanted boxes of matches or other small goods at her in lieu of the required small change.

My good lady wife, of course, shouldn't really have been out doing the shopping at all. We, like everyone else of our 'standing' and culture, already had a maid for all of that.

Ellie was far from happy about this, and it wasn't something which I'd ever expected to need, either; not when residing in my own home along with my own non-working wife. But it was the 'done' thing, and we were therefore told that we had to conform; proper uniforms and all. Our flat even had a maid's quarters: a concrete-walled side room for them not just to work, but to cook, eat, and sleep in, if they were so required. Both of our maids, thankfully, had their own homes and families to return to at the end of the working day: I don't think that we could have coped otherwise.

I first engaged a local woman recommended by the couple in the flat below. It was not a success. Julia spoke only Brasilian, and sometimes quite brusequely at that, so Eleanor couldn't therefore convey anything more than the most basic requests and instructions. Come each evening, we also seemed to be missing a lot of food.

Having pretty quickly dispensed with her services, we asked around a bit more, and then took on Maria, a slight and very timid young woman, who seemed ever so grateful for the job. She and Eleanor got on much better, although I don't think that Maria ever really recovered from the shock of finding out that this 'rich' foreign lady, surely with many other leisure interests and social engagements to occupy her time, really didn't want that much doing around her home at all - although relief from the chores of manual washing, dusting, and ironing was generally welcomed - and rather liked attending to her own food shopping, English-style cooking, sewing, mending, and various other household matters in her own established ways. So our lucky maid - who we also decided was grossly underpaid, albeit fully in accordance with the prevailing rate - spent a good amount of her day sitting in her quarters, eating her staple diet of brown beans and rice, attending to her own sewing, and leafing through various magazines. She was, to say the least, very sorry to see us go when we eventually returned back to England.

I had my own car and chauffeur, too; well, kind of. The provision of a driver with local knowledge and language skills wasn't, of course, an unusual occurrence whilst out on field or site survey work: but it felt strange when I both worked and lived in a residential city and office-based setting. Still, there was an argument that it was a business necessity: as well as our frequent flying visits to various companies, ministries, and institutions in Rio and São Paulo, Luca and I had by now set about in earnest visiting as many sugar factories as we could in Pernambuco and its 'nearby' neighbouring states of Alagoas, Bahia, and others. These trips, promoting our new company's services, often involved very long journeys by road, and either overnight stays or late night returns. I wasn't going to complain if I didn't have to drive.

Young Roberto seemed quite competent and safe - as far as Brasilian drivers went, anyway - and he certainly looked most handsome in his grey uniform, complete with black-peaked cap. He also lived in quite an impoverished quarter of the city, if only about a mile away from me, and there wasn't any parking for the car at his home; it wouldn't have been very safe there in any case. So we arranged that he'd leave it each night at my flat's private facilities, and return on foot the next morning to drive me the three or so miles across the wide river, past the steaming, brightly-signed Bacardi distillery, to the far side of the city centre where our offices lay.

I say 'car', but the vehicle itself was actually a huge Chrysler pick-up-cum-estate, a petrol guzzler far larger than we really required for any purpose. According to my Brasilian counterparts, however, it gave a certain "dignity" to Cosinor's operations. Either way, since it had to be left with me for safekeeping, once I'd finally obtained my official Brasilian driving licence - something, for some reason, which not even Luca could manage to hasten along - I thought that I might as well use it to undertake some evening and weekend trips of our own.

Recifean driving, I have to say, was most definitely neither a dignified nor safe experience. It was, frankly, to the uninitiated and initially unpractised, simply terrifying, the accepted local standard of driving extending far beyond the normal definition of reckless, not to mention aggressive. Passing on the wrong side of the vehicle was the norm - preferably with someone else on the other side at the same time, too - and beating the lights almost as favoured a national sport as their beloved football. They also liked to make up their own versions of the Highway Code, including such permissible time- and distance-saving actions as travelling the wrong way up a one-way street "just to get to the next intersection, you'll understand".

Collisions - also called Batidas, generally meaning something in the region of a shake, battering, or beating; and as such explaining quite a lot about the name of that particular drink - were, accordingly, all too common, and often involved significantly-injured, and far too many fatal, casualties. One of the most favoured places for such an incident was encountered upon leaving the city centre and joining the southern coastal road, where a five-lane line-up at the traffic lights led directly - and I do mean directly - into a two-lane stretch of subsequent road. It was a trip which I foolishly made at least twice a day. When the lights were at red, the whole five lanes quickly filled up with revving engines, all keen to make the fastest possible Grand Prix style getaway at the first glowing hint of green. All Brasilians, of course, revered their famous national racing drivers; clearly, everyone wanted to drive like them, too. Now, I began to comprehend how they trained them.

The roads in Brasil were otherwise, to be fair, remarkably good, and afforded a good many other options to the country-wide traveller, should he or she not care to fly the oft-terrific distances involved. Luxurious sleeper-seat coaches, for instance, plied their frequent way between Recife and São Paulo, and all stops in-between, taking two continuous days to do it. Within the cities, however, if not the north east in general, congestion simply ruled. As much as anything, this was simply due to the roads' hugely-diverse variety, and therefore speed, of vehicles: the horse-drawn and larger, wooden-wheeled, ox-hitched carts, often overflowing with recently-cut cane; the assorted cars, vans, coaches, and busses; the huge goods lorries and cane-haulage trucks, all far-from-securely loaded to well beyond their rated capacities.

From the very first day that I'd stepped off the plane, I was struck - metaphorically, thankfully; if only just - by the preponderance of Volkswagen Beetles upon these crowded streets. Most, it turned out, were taxis: but there were many private ones, too. The European production of that particular iconic marque of car had by this time just about ceased: but in São Paulo, they were still being manufactured by the thousands. Small Fords were also quite popular, perhaps making up about one in twenty of the overall vehicle population: although, whilst the bonnet and badge seemed quite familiar, they somehow appeared oddly different from those seen in the UK. I eventually learned that they were also being made in Brasil, but to a new, local design combining the Ford Escort with the French Renault Dauphine. Well, of course; why not?

There were some larger cars around, too - many, like ours, US imports - a good number of which formed a particular type of public transport known as a 'lotosa'. These long, six-seater cars operated on fixed routes from the city centre to a particular suburb, and could be flagged down, if they had a spare seat, at any point along the route. The passenger paid the driver a simple, fixed, upfront fare, irrespective of the length of their journey, and then disembarked upon request at their preferred place. It was, in effect, a reasonably efficient and affordable private bus route, with the added advantage of a guaranteed seat: a rather different experience to the regular public buses, which, as in so many 'third world' countries, had to offer extremely cheap fares and, in consequence, were always crowded to bursting.

The busses didn't look too roadworthy, either: but then again, neither did many of the Beetle taxis. I regularly travelled in ones which had their rear doors completely removed, whether deliberately or as the result of one or more impacts I never knew. Still, at least there was a kind of welcome ventilation in the stifling humidity, and it certainly made entry and exit all the more easy; as long as one clung on in the back, unstrapped and unbelted, in the meantime.

Oddly, and not in the least ever assisting the general chaos, there was one road traffic regulation which all Recifean drivers did seem to obey; took very literally, in fact. This was the one which made it obligatory for every vehicle to remain exactly where it was whenever any accident took place. I'm sure that it had much less to do with any lingering regard for the law, and much more to do with satisfying their ghoulish curiosities, especially if blood was spilt. Nevertheless, at the scene of any collision, no matter how minor or major, the traffic flow could easily come to a complete standstill for an hour or more, awaiting the eventual arrival of the police, the taking of the appropriate notes, and their final permission for the resultant tangle once more to attempt to proceed. Accordingly, even when dealing with the most routine, and hopefully short and swift of journeys, such regular delays always had to be planned in and allowed for, as just one more part of the overall concept of Brasilian time.

Still, it wasn't all bad, and having the use of our own car meant that we could visit many nearby places of interest in our spare time. One of our first out-of-city ventures, whilst I was still learning to drive in the true Brasilian style, was the indeed-beautiful little town of Olinda, just up the coast, where there still stood some lovely old, colourfully-painted buildings, including a typical twin-towered, central-American-styled church. Not so beautiful, but equally fascinating in its own way, was the former slave market, its remaining cells surrounding a square whose centre contained a raised platform and a small podium, from which the slaves had been exhibited and auctioned off.

We made a number of other journeys to the 'interior', as almost anywhere outside the thirty mile coastal strip of land was called. The more seawards areas were, for the most part, extremely hilly and broken-up, almost like a miniaturised Swiss landscape: except that the steep, red-soiled sides here were all covered in fields of sugar cane, often seen being carried down on the backs of mules, and subsequently transfered into either wooden ox-drawn carts or large, diesel-driven trucks.

Further inland, the country became flatter and much more arid, almost like a desert. Right in the middle of this 'Sertão' scrub country, about ninety miles away, we found Novo Jerusalem, a fair-sized modern town complete with shops, restaurants, and hotels. Its main purpose of existence, however, as the name might suggest, was the staging of an annual Passiontide play. Dotted around a very large area of grounds, various permanent structures each represented the main buildings in Jerusalem where Christ was said to have spent his final days, including the house of the Last Supper, the Temple, Herod's and Pilate's palaces, and the hill of the crucifixion. Whenever the play was performed, the audience, instead of being seated before one single stage, walked between these various places scene by scene, watching as events unfolded. Visitors and pilgrims from all over Brasil apparently flocked there in their hoards at Easter. We, of course, went at completely the wrong time: but we were still able to wander around, and used our imagination.

Our travels soon found us exploring several of the more remote villages along the nearby coast, too. Their idyllic, palm-strewn beaches were straight out of Desert Island Discs; and often deserted to boot. One day, whilst having our usual picnic, a group of young boys approached us from out of the trees which lined the back of the sands. They were holding out for sale bunches of plump, bright red and yellow fruits, looking rather like the sweet peppers which we're now used to back in the UK. But these, I knew - if from nothing but the ubiquitous paintings, ornaments, carvings, and mutifariously-useful shaped wooden and clay jars which one found at every Brasilian turn - were cashew fruit, their small, curled, stalk-like nuts, unripe at this stage, still hanging from their undersides. The local people, so I'd heard, not only harvested and dried the ripened nuts, but made a palatable drink from the fruits' juices: so we bought some, and took them back to the flat. What I hadn't heard, and didn't know until I took a bite out of one of them, was that not only were the fruits themselves wholly inedible, but that their pure, unadulterated and undiluted juices were extremely acidic. Within a few minutes, I had blistering burns all over and around my lips, deep enough - not to mention extremely painful - to leave permanent scars.

It was beside one of these backcountry roads that we found one of our favourite places to eat: a sizeable wooden, open-terraced shack in the middle of not much else. I can't remember what its proper name was, though, because we all knew it simply as 'the blue lobster place': probably because all along the length of one wide inner wall was a huge glass-sided, water-filled tank which contained a good number of beautiful blue-shelled lobsters. Whether they were originally caught somewhere or just bred on-site, I really don't know - I have some recollection that there were a number of other large, sunken tanks in the property's back garden - but their indoor enclosure always seemed to be very well-stocked.

Although this somewhat rough-and-ready restaurant was clearly very popular with folks from all walks of life, you never had to make a reservation: you simply arrived and took a seat wherever you could amongst its solid, stained tables and benches, sometimes just squeezed in alongside everyone else. When the waitress came along and asked what you wanted, it wasn't really a difficult choice, because "lobster, please" was the only possible answer. Not even "I'll have lobster with...", because there wasn't any 'with', either. The only other questions left to answer were "how many?", and "which ones?": for this, just like we were used to with summer fruits in England, was also - allegedly, at least, because sometimes I do wonder if it wasn't all some kind of enterprising marketing gimmick - a 'pick your own lobster' joint. You rose from your seat, took the lady over to the tank, and pointed out your choice (not unlike, I had to recollect a few years later, the girl dance partners in Thailand; although hopefully, the outcomes were at least a little better for them); at which juncture the doomed creature was skilfully extracted, then taken off to the thankfully-behind-the-scenes kitchen. With its soul dispatched off to The Great Sea In The Sky, the warm mortal remains were soon presented back to you upon a well-worn wooden board, along with a large napkin, a bowl of water, and a chunky wooden mallet.

And when I say its remains, they certainly did want to remain firmly within their blue-tinged shell, for, to get at the delicious pinky-white flesh underneath, you first had to hammer vigorously upon, and then pick and rip off the pieces of, its cracked outer casing, eventually digging your fingers around and into what was once its claws and other internals until you could get out every last tasty morsel. Hence the water and napkin.

We didn't always have to travel far to find something of interest, though. Nor, of course, to find a magnificent beach, for we had one right on our own doorstep - once the perils of crossing the intervening road-cum-racetrack were successfully overcome - extending all the way southwards from the city's outskirts for at least another five or more miles beyond where we lived. This long, thin expanse of gently-sloping sand was washed repeatedly by the sea, and thus was not only very clean, but excellent for beachcombing. There was also, about a hundred yards out, a narrow, flat-topped, stone reef - for which the Brasilian word is recife - often only noticeable by the spray breaking over it, but usually becoming exposed at low tides. At those times, being ever mindful of the occasional much deeper pools, it was then possible to wade out across the intervening clear, calm lagoon in mostly, for a grown adult, only waist-high, wonderfully-warm water. This sheltered stretch also, of course, made it an ideal area for all ages to swim and play. The main beach, consequently, sometimes became very crowded, especially at weekends. But we, living so near, were lucky enough to be able to visit at much less popular times. I taught Rai to swim there, and later, after some pestering, we bought her a little polystyrene surfboard, so that she could try to ride in some of the gentler waves which came over the reef.

About half a mile further along from our flat, heading away from the city, was the Praça (square) Boa Viagem. At the time, it was the end of the main beach-front road - not too long ago, it had been an entirely separate outlying village of the now ever-expanding city - and thus a natural termination point for our regular strolls out along the sand and back. Surrounded on three sides by shops set into low buildings, its grassed and paved areas were shaded by large palm and fig trees, and dominated by two stand-out features: to the seawards side stood a reasonably modern, cube-shaped hotel block, about ten storeys high, whilst at its centre there was a large, old, typical South American church, dedicated to Our Lady of Good Voyages. It was a known centre for tourists, as well a gathering point for local friends and families, the dusty pavements always populated by ragged men selling various trinkets and mementos, some of which they made on the spot. We often bought balsawood toys for Rai - wiggly, multi-segmented, wire-jointed snakes and head-nodding turtles were favourites, together with little hollowed-out penny whistles - as well as both decorative and useful items for ourselves: wooden cashew jars and model boats, clay figurines, hairy cow-hide shoulder bags, and some surprisingly-comfortable collapsible, three-legged stools, upon whose triangular, thick leather seat one sat like a pommelled saddle. (Both Rai and Faye, I know, still have and use theirs, some fifty plus years on, and haven't ever needed to repair them.)

All along the top side of the main beach, where the road still ran alongside, there was a long line of palm trees, underneath which were set staggered, individual stalls, variously selling a selection of sweets, bottled drinks, and vanilla or coconut ice creams. There were also those which sold fresh coconut milk drinks: and when I say fresh, I mean it, these vendors' locations always marked by a massed pile of light green, football-sized fruits - their smooth, shiny skins about as far as you could get from the small, brown, rough-fibred husks which we'd come to know in the UK, but exactly the same as those currently ripening on the trees above your head there - a man brandishing a broad, sharp machete, and a packet of colourful paper straws. For a few Cents, the man would reach down and select a coconut from his stack, and then, with one swift stroke, slice off an end, insert a straw, and hand the whole thing over. The cool, sharp juice inside - actually clear, despite its name - was very refreshing indeed, and the top of the beach, for a good distance around, was always strewn with slowly-rotting outer shells and discarded straws.

Business on the sands themselves was also usually brisk; if not sometimes frantic. Sheltering under large sun umbrellas, each little group of beach-goers would be approached every few minutes by one of the many itinerant salesman, their cries preceding them, offering just about anything which one might need: drinks, ice creams, confectioneries, toys, buckets and spades, even clothing and household equipment. This beach trade was really only an extension of the wandering street vendors, men who continually plied the local roads at all hours of the day, and often night, loudly shouting out their wares so as to be heard within the buildings and houses. They would mostly be purveyors of various foodstuffs - oranges and other fruits, biscuits, vegetables, fish, cakes - but, you name it, from metal implements and pots and pans to coat hangers and clothing, it probably passed by at some time or other. Each seller had his own individual cry: "Há camarão, abacaxi" (shrimp and pineapples), and so on. One of the most common was the "Há peixe" man. He came around with large fish - I'm talking several feet long, in some cases - suspended by hooks and string from a pole slung across his shoulders. If stopped by a purchaser, he would remove the requested fish, put it down upon the pavement, and, having asked how much was wanted, proceed to take a long knife and chop off the required section. This was then weighed on a hand scale before being wrapped in a piece of old newspaper, usually carried tucked under his belt. It wasn't exactly the most hygienic of set-ups to start with; it certainly became even less so later in the day, when the flesh had been exposed to both heat and flies.

We continued to buy our supplies from Comprabem.

Still, at least the fish were locally sourced, the men selling them also being those who'd earlier caught them from their 'jangadas'. These fishing boats were, in reality, more like a one-man, sea-going raft: about five feet in width, but much more in length, they carried a single, angled mast, a large triangular sail, supported by a jib, and a small seat and long steering oar fixed at the back. The deck was constructed by lashing together several logs of some type of very lightweight wood, so that the vessel, although quite sizeable, could easily be lifted into and dragged back out from the water without further assistance. The fishermen used only lines for their catches, yet they went out considerable distances, knowing all of the reef's gaps and low points through which they could access the deep waters beyond. Consequently, variously-sized groups of jangadas would always be found drawn up upon the beach wherever there was either such a gap, or otherwise sufficient depth of water at high tide to sail right over.

This very basic, subsistence-level existence was a common, widespread, and oft-repeated theme during my time in Brasil. Whether travelling on business or for pleasure, so much of the country which we encountered was markedly poor, not just in an agricultural and cropping-capacity sense, but in terms of its people's lack and hunger, and the everyday difficulties of their lives. Even on the better sugar estates, whilst the affluent owners and managers enjoyed their days in ostentatious luxury, the workers still lived in very sparse, rudimentary conditions.

The truth of the situation was brought home to me one day when, off on one of our long road journeys, Luca, Roberto, and I stopped for lunch at a roadside restaurant. We sat out in a paved and planted courtyard which was surrounded by a high stone wall, in turn topped with barbed wire. This was, I presumed, for night security: but I quickly found out that it also fulfilled another purpose, for, as we ate, I began to hear children's voices behind me. They must, I thought, be playing nearby.

Then Luca said offhandedly: "Take no notice."

I didn't understand at first. But then, on turning my head, I saw a dozen or so small, filthily-dressed boys clinging to the top of the parapets, shouting and gesticulating to us to throw them some of our food. Upon a concerned enquiry, I was again instructed in no uncertain terms that we should not encourage such behaviour, and that the "peasant children" must be driven away, so as not to bother visitors to this establishment, thus giving it a bad name.

Such superior, elitist attitudes were commonly found within the managerial and more affluent classes; sometimes - perhaps often - even from my close and well-regarded colleagues and friends. They were typified by another family with whom I stayed on several occasions, and eventually got to know quite well. Paul Quelt was an American citizen who had married a sugar estate owner, one Dona Camilla Da Santa Oliveira. She, being the only child of an old Portuguese aristocratic family, had inherited sole ownership of their factory and lands, the latter including a township which housed the workers. There was always intense rivalry between neighbouring estates, so Dona Camilla had initially sought to overcome this in a rather traditional, old-fashioned way: she had married the adjacent property-owner's son. Unfortunately, antipathy between the two families had not only continued, but exponentially grown. Somehow, the Brasilian Roman Catholic church had agreed to allow her a divorce, and she had then had married Paul; an occasion which, of course, only served to deepen the old resentments even further. One assumes that it can only have been true love; or the fact that Paul was president of a very substantial Recife-based bank, and significantly wealthy to boot.

Naturally, Paul and Dona Camilla lived in a magnificent house, surrounded by landscaped grounds which included an Olympic-size pool. As a family, we were invited out there several times for weekend stays, where Rai enjoyed rides on their daughter's prize pony. What we didn't tell Paul or Dona Camilla, though - and they never said anything, either, although I'm pretty sure that they must have known - was that Ellie and I were also quite often invited over to social events 'next door'. These usually became very late night sessions, were of an altogether different nature, wholly unsuitable for a young child, and without a horse in sight. But there was always a Luca in tow: and he, of course, always insisted that, even though it wasn't technically in any way a business occasion, it was still properly and necessarily incumbent upon us all to travel in style. There was therefore also always a Roberto to drive us to and fro.

Our personal experiences of both of these sets of occasions - apart from the standard thoughts of 'how the other half live', as we reluctantly returned to our city flat - were never anything but pleasant and most congenial. I once spent a few days at Dona Camilla's estate alone, however; or, more correctly, along only with a colleague who'd just come out from England. And then, certain things suddenly started to look a lot more clear. One day, for instance, whilst happily enjoying the estate's amenities after our day's work, we noticed a large avocado pear tree, growing not at all that far from the house. It was currently bearing a wonderful crop of fat, ripening fruit, and we happened to comment on it to Paul over evening drinks, adding - tinged with only a little jealousy - that it must be nice to have such an appetising and, to us at the time, rare and luxurious food so readily available. We were, of course, hoping that he might also serve us up one or two.

"Oh," Paul shrugged dismissively. "Those ugly things? We never touch 'em. They're just left for the garden boys."

We couldn't help but to express our surprise, explaining what we thought that he was missing. So Paul summoned a gardener, and instructed him to climb the tree and bring back the largest ripe fruit. With this duly delivered, along with a knife and spoon, Paul proceeded to taste a sample. It was immediately declared to be quite gloriously delicious; and the garden workers were henceforth, forthwith, and so on, forbidden ever to take any more for themselves.

It wasn't quite what we'd intended to happen. What's more, he sat there, right in front of us, and ate that one, single, succulent avacado without even offering us the smallest slice. Neither did he ever think to repay the favour at any later time. It reminded me of exactly the same disappointment which I'd felt as a young boy when visiting my Uncle Charlie, the head gardener at a large stately home in Leeds, and not being allowed to try even one of the rows upon rows of ripe, plump, sumptuous-looking, cordon-trained peaches which he so dutifully tended. No: such delicacies were only for the people at that big house, too.

In a similar vein, when we told Dona Camilla how impressed we were with some of the root vegetables which we'd seen growing in their grounds, she admitted to never having bothered tasting the likes of yams and manioc (cassava). They were also foods only for "the poor people". Like her husband, however, she did condescend to send the cook to fetch some, just to humour us: and, of course, as soon as she'd had them prepared and presented to her, she likewise realised what she'd been missing, and resolved immediately to remedy the situation as regarded her own needs.

Still, Dona Camilla did occasionally show some concern for her workers. It wasn't much, probably nowhere near enough in the grander scheme of things, and they still lived in the exact same poverty as other labourers: but she was, at least, prepared to provide them with some basic medical attention, should they fall sick or be seriously injured. It was, to my knowledge, a great deal more than many of her contemporaries were willing to do.

Perhaps, though, some general glimmers of hope did exist. Soon after I arrived, Luca suggested that I should go with him to a weekly lunch meeting called 'Lideranças Christão' (Christian Leaders).

Held every Monday in a central Recife hotel, this group brought together all types of businessmen, teachers, university professors, and similar 'senior' city residents. Some were political, some not, and their common beliefs ranged widely in both denomination and attitude, from the very formal to very free-thinking; Roman Catholics, a variety of Anglicans, non-conformists of all kinds, and even a few Evangelicals and Charismatics. Yet, just as with the school's ecumenical church, there was nothing here but a shared desire to sink whatever divisions might otherwise exist between them, and to exchange ideas about, and then promote, whatever good works they could for the assistance of a community within which so much dire need existed.

Whenever the group met, there was, before the commencement of a simple meal, a short prayer session, followed later by a bible reading; all in Portuguese, of course. Everyone was expected to make some contribution, if just by adding a few words to the post-prandial discussion. Even at my very first meeting, I was asked to try to read out at least a few verses from the 'Novo Testamento' in my still very halting attempts at their language: an encouragement for which I was later very grateful, as, soon acquiring my own copy, I was then able to practice my reading, and assist my linguistic learning considerably with the benefit of a work to which to my ears and mind were already familiar.

I continued to attend these meetings right up until the time when we left Brasil for good. By then, I could join in all of the conversations with relative ease. But these sessions, together with the EAR gatherings, helped me to acquire far more than just some words and grammar, heavily influencing, and in some cases significantly changing, both my philosophical and spiritual thinking to a level which remains with me to the present day. They also served to bring me into contact with many people whom I otherwise couldn't possibly have met, across and within many walks and positions of life, and showed me that there were indeed some in Brasil who remained, to a degree at least, genuinely concerned about the state and futures of their country's peoples, and not just wholly in promoting their own positions and wealth.

To be honest, I never could fully get to grips with the Brasilians' whole general mix of attitudes back then, whether towards their fellow citizens' living conditions or otherwise. The whole place was a living, oddly-balanced, contradiction. As a people, they were extremely proud of their country's history and achievements; not to mention its football prowess. They valued and boasted of their antiquity; yet, in the next breath, would be extolling the virtues of their incomparable modernity. Recife city was a prime example, poverty and opulent luxury often as not sitting side by side in terms of its buildings; the favela shacks and the shining skyscrapers; steaming huge factories and tiny folk-craft workshops; ancient churches amidst towering hotels and offices; the beggars and shabby street vendors in permanent residence outside modern, brightly-lit department stores; sleek, air-conditioned, long-distance coaches being passed on all sides by smoke-belching Beetles.

Yet overall, oft as not, it still came to nothing when presented with the final test of putting ideas into practice. At each meeting which I attended, from every person whom I met, I encountered nothing but an air of unbounded enthusiasm for every possible thing.

"But we do not need to change or develop," I was then regularly informed.

This latter claim did seem rather contrary to the second part of the 'Ordem e Progresso' motto emblazoned across their national flag; and perhaps, in many aspects, one ought at this juncture simply to forget certain possible interpretations of the former word, too.

"We have the oldest sugar industry in the world," the declaration usually, in my case, followed: a fact clearly evidenced by the antiquity of the systems and equipment employed in many of the small mills and factories which I visited, some so old that they were practically working museums. "We have grown cane here for over four hundred years."

"Yes," I felt like adding: "And you still had slaves until 1888."

Given the attitudes of some of the estate owners, I did sometimes wonder if that last date had ever actually been passed.

Of course, new ideas and ways, as that minister had reasoned during my first visit, weren't always so ideal, either. To counterbalance the ageing facilities, I saw many of the very latest designs: again, often as not, right next door. Several new manufacturing and processing sites had recently been set up in the north east, especially within the textile industry. Much of their machinery had been purchased from England, too. It should, at the least, have made me proud, never mind quite hopeful in a business sense. It didn't, however, always turn out to be such a good thing, for, in a territory teeming with eager, unemployed labour, these almost fully-automated plants required only the absolute minimum of workers to operate them.

So much, indeed, for progress.

There was one short trip which, for me, summed up the whole, typical problem more than anything else. Not too many years earlier, the government had implemented a scheme to colonise and improve the agricultural and other resource-utilisation capacities of the country's northern Amazon basin. They were now constructing a new road, stretching from the populated north-eastern states, through the river territories, over to the northern Brasilian border; and then, ultimately, connecting into the Central American lands. This Trans-Amazonian (or Amazonica) Highway had, whilst I was there, just passed the town of Altamira, lying on the southerly Xingu tributary about halfway between Belem, the main port of the Amazonian delta regions, and Manaus, a large inland city set deep in the heart of the rainforest, along the great river itself. In addition to the new road, many farmers and peasant families were to be - and already had been - resettled at sites spaced every few kilometres along its route, these being created by the expedient felling of swathes of trees, their stumps simply left in situ. Each person or family had been given, free and gratis, several hectares of land on which to build their hut-cum-farmstead, together with enough corn seed and fertiliser to see them through their first year, growing wherever and whatever they chose. Thereafter, it was up to them to do the best that they could.

Great things were, naturally, expected.

The government had now come up with plans to extend the country's sugar industry into these remote, developing 'new territories', too. Luca was accordingly keen that we should visit the proposed cane-growing area as soon as possible, in order to make an early offer for the supply and construction of its allied factory and plant. The site was only a little way north of Altamira, along and taking beneficial advantage of the expanding highway, and so arrangements were made for us to fly up there early one morning, via Belem. Our journey to the nearby, recently-constructed airstrip was, as previously recounted, a little more than fraught. Once safely redeposited upon terra firma, however, we clambered with relief into our pre-arranged hired Jeep for the relatively short, thirty-to-forty mile drive along the grand new road.

'Road', as we soon found out, was a somewhat erroneous, certainly over-optimistic, description, it then being - and, I believe, mostly still is - little more than a wide dirt track, running straight as a sometimes significantly-undulating die between the tall, densely-packed trees on either side. The surface in some places was hard, dusty, and seriously rutted; in others, not in the least assisted by the helpful process of continually grading it back to smoothness, nor some recent heavy rains, it formed nothing but an endless, and far-from-amusing, skid pan of slippery, orangy-red clay.

We passed many of the new 'villa' sites along our way. Some were even observed from the perspective of a generally frontwards-pointing vehicle. Very few still seemed to be occupied. They obviously once had been: but most were now clearly abandoned, their huts and other small buildings already in ruins. There were none of the planned crops growing, either, and the ground between the tree stumps was visibly devoid of soil. Pressing onwards into areas where the trees had been even more roughly felled, we finally found some signs of habitation: but even here, the crops looked terribly poor. It wasn't very difficult for me, at least, to understand. All that had ever existed between the original forest trees had been a very thin soil, covered by a layer of fallen leaf. This topmost level, in the great cycle of nature, had provided all of the trees' necessary nutrition, all by itself: but as soon as the trees and their leaves had been removed, and the soils in-between exposed - and never forgetting the area's heavy seasonal rainfall - each site held a sustainable growing lifespan of about a year, maybe two at a maximum, before it was effectively fully depleted, eroded, and then finally physically washed away. The poor settlers, of course - already financially poor before they'd even been relocated, never mind whatever sympathy one might now carry for them - were left with nothing but dashed and deluded hopes of their promised, prosperous futures, and had either deserted the area completely, or simply moved up the road to start a new site for themselves in an unauthorised location; until it, too, became unusable, and they continued their new, nomadic life some place anew.

Our current visit also had a very limited lifespan, for we'd been told in no uncertain terms that we had to be back at the Altamira strip in time to complete the first leg of our return flight well before dark: a stipulation with which, even given the assistance of the new automatic direction finder apparently just most helpfully installed at Belem, we were only too keen to concur. After mile after mile of time-eating slipping and sliding, therefore, we were growing more and more sure that we'd never be able to reach the soon-to-be cane fields.

And then, suddenly, it was announced that we had arrived. There wasn't exactly much to see for our troubles - not that I was actually expecting it yet, in any case - just more of the same tall trees, and, amongst them, a minimal rectangular patch which had been completely cleared, to the extent that there was, yet again, absolutely nothing left but barren, soil-stripped ground. How anyone would ever be able to grow cane in that area baffled me completely: but then, "This is Brasil!" would doubtless from somewhere come the confident cry. Someone had so planned it - in the far-away, architect-designed, modernist, landscaped, federal capital of Brasilia, like as not - using only a map and maybe the odd photographic print, without any knowledge of the terrain, climatology or soils, and without first asking the right questions of anyone potentially armed with that knowledge; and so, just like the planned smallholding sites, thus it would duly be.

Or not. Altogether, it looked and felt like nothing but a thoroughly wasted, and rather saddening experience. It certainly wasn't something which either Luca or I wanted any further part of, so we immediately turned around, and skidded and spun our way back to catch our - thankfully wholly without incident - return flight.

Talycopi wasn't, by far, the only outside interest in which Luca played a significant part; and, by association, with which I therefore also got involved. Brasil's gemstone industry was already considerable, the Sterns dealership controlling most of the mining areas. But Luca had heard that, deep within the remoter parts of the inland state of Minas Gerais, further down towards Rio and São Paulo, there remained huge untapped sources.

It so happened that an English entrepreneur, Hugh Storey, had recently come out for a stay in Recife. For some reason or other, he'd been directed to make contact with Cosinor, where our office also lay. This, of course, presented a golden opportunity for Luca to promote the great riches just waiting to be gained through the exploitation and exportation of Brasilian gems: if only someone could be sent out to discover their whereabouts. Being the main English-speaker, I was inevitably drawn into the discussions, and thus became party to the highly-secretive plans which were subsequently drawn up.

One morning, I was unexpectedly asked to accompany Roberto out to the airport. Someone apparently needed to go and welcome the English geologist who'd just been urgently engaged to spend time down in Minas Gerais, investigating these very such possibilities. I was obviously thought the most suitable, or perhaps just most easily-spared, for this important task. It was certainly a task which took up an unexpectedly large amount of my time, due mainly to the unusually large amount of Brasilian-style bargaining required - without a single EdG card or badge, never mind a fluent native speaker, to ease it along - in order to retrive our new visitor's geological hammer. Somehow, it transpired, he'd been allowed to carry it upon his person all the way through his departure airport's security, and right up until the point when he'd boarded the plane. The crew, not unreasonably, had then decided to relieve him of it. The pilot, upon subsequent landing and disembarkation, now wasn't minded to give it back.

Having chased and negotiated our way around most of the airport several times - trying, as much as anything, to track down said pilot before he flew off again on the next leg of his routing - we finally managed to transport our man, complete with hammer, over to his hotel, and thence to talks at Cosinor's office. He set off on his survey the very next day, and returned several weeks later to report to Mr Storey that there were indeed plenty of gems to be found, but that a more widespread study was needed in order to ascertain the viability of their extraction. As a result of this, therefore, two British postgraduate geology students were then swiftly employed, and also dispatched to Recife post haste, in order to find a site where such a plant could be set up.

The area in question was, even for Brasil, particularly remote and undeveloped, with very deprived living conditions and effectively non-existent communications. Storey's instructions, moreover, were that their work was to be carried out with minimal expense. The pair was therefore sent forth by the initial means of the long-distance, forty-eight hour coach service, and then expected to make their own way for the remaining several more hours required to reach their final destination.

They were never heard from again. Well, for at least four weeks, when, completely unannounced, the two of them turned up at our office one afternoon, looking very dishevelled and dirty. They were very pleased to confirm the availability of many and various stones, they stated wearily. Sadly, they weren't so pleased to have to add that the gems' varying sources were scattered over some considerably-wide areas, none of which were in sufficiently close proximity to be able to establish any central, commercially-justifiable collection area.

Thus, then, ended Luca's dream of a new, or at least large-scale, Klondike. Whether anyone else ever went back and exploited the area on a smaller scale, I never knew. We had the students round to our flat for a meal after they'd cleaned themselves up, had a very long and interesting chat about conditions in the more central reaches of the country, and then saw them off at the airport the next day, all hammers safely in suitcases, for their return to the UK.

Although happily settled into Recifian life, we decided to spend Christmas back in the UK, alongside our other three daughters and their own growing clans. Annoyingly, therefore, because of a recent Cholera outbreak in Portugal, even though we'd only have to transit through Lisbon airport on the way home, the international health authorities required us all to obtain further vaccinations before being allowed to travel. We'd had the injections before, of course; just a quick, gentle jab into the buttock muscles. It wasn't any major issue.

Excepting that this was Brasil. Presenting ourselves as directed at the medical department of the city's police headquarters - no, I don't know why it couldn't have been done by a hospital clinic or local nurse or doctor, either - Eleanor, Rai, and I were shown into a large, wood-floored room, barren except for a number of old, desk-like tables, unscreened and fully open to view, where several men were clearly in the midst of both the giving and receiving ends of its temporarily-designated purpose. We were then told, quite unceremoniously, to remove or otherwise adjust our clothing sufficiently enough to expose our bellies. Apparently, the doctor adamantly stated, after I had the temerity to ask why they didn't favour the more usual anatomical site, or perhaps an arm or leg muscle instead, everyone knew that the upper abdomen wall was a much better - if, to us, also much more embarrassing - route of administration. It was certainly a lot more painful, both immediately and afterwards, our sore and stiffened stomach muscles meaning that none of us could properly sit down for about a week.

No wonder so many people were scared of healthcare in those days.

Arranging our home leave was otherwise problem free; our journey much less so. It wasn't a trip which we'd planned too far in advance, so there was no surprise when we found ourselves not exactly spoilt for a choice of flights. The only suitable seats left available, in fact, meant landing back in England just two days before Christmas Eve. This was cutting things a lot closer than experience told me was wise: but it was a frequent, scheduled route, I reasoned, and a major, and reasonably-reliable carrier. Everything should be fine.

Which it was: until, upon approaching the European coastline after our overnight, trans-Atlantic crossing, we were told that Lisbon airport was fogbound, and we were being diverted to Oporto, a good distance north. Once there on the ground, we were further informed that our plane's flight crew had already exceeded their day's allotted air time, so they couldn't then fly us onwards, even though Lisbon had by now cleared. There also weren't any other local flights or spare planes available - this wasn't, after all, Brasil - nor did anyone fly out from Oporto to the UK. There was nothing which anyone could do.

And there, the information ceased.

We, along with the numerous intermingled passengers from three other large, diverted planes, managed to kill a not entirely amusing couple of hours dealing with the ensuing baggage-handling carnage; but then, still with no assistance forthcoming, with no-one helpful even in sight, everyone just stood around and waited, and waited, without any other form of entertainment, refreshment, or other meaningful comforts.

About four hours later, someone eventually appeared and announced that we would all be taken to Lisbon by coach.

It was well after dark when we finally arrived, still unrefreshed, after a long and uncomfortable journey. Our planned onwards flight home had, of course, departed hours before. Yet again, no-one else much seemed to care, and no-one had any alternative suggestions. In the end, I managed to find three seats back to Gatwick with another airline, leaving at ten the next night. With a one-hour time difference, I hoped that this would allow us just enough time after landing to get the last train back to our local station - if not, it'd be a long, cold wait for the first train of the next morning - and then a short taxi ride home.

The flight, of course, arrived late. Unbeknown to us, however, our daughter Faye and her husband had already decided to come down early from their Midlands home, and thus were presently waiting for us back at our unexpectedly-empty house. By some miracle of then-modern telephony, they'd already checked with the airline for delays to our original schedule, found out about our plight, and somehow, in a way which these days wouldn't be at all permitted due to several sorts of data and privacy laws, eventually managed to locate us on our rearranged itinerary; and then, very thoughtfully, driven out to the airport to meet us.

We awoke from our exhausted slumbers the next day, Christmas Eve, and, leaving Faye and Chris to handle the arrival and accommodation of the others, set out with urgency into Croydon to do our outstanding shopping. The shelves in all of the town's main stores were virtually empty. But no matter: we'd made it home in time to enjoy our traditional family celebrations just the same.

It always paid to double check all possible Brasilian rules and regulations, especially before attempting to do anything as adventurous as leaving the country whilst intending to re-enter again. Wanting everything to go as smoothly as possible, therefore, I'd already made appropriate enquiries well in advance of our Christmas trip.

But we already held full six-month visas, came back the astonished reply. They wouldn't need renewing until after we'd got back. So, no, we wouldn't need formal departure permits. No, we wouldn't need permits for our re-arrival. No, there was nothing else required as regarded our scheduled new year's return and continuing residence.

It couldn't really be that simple, could it?

They did, to be fair, still allow us to re-enter the country; if not for very long. But we couldn't now, apparently, in complete contradiction to the assurances only so recently given by that very same office, if not probably by the exact same officials, obtain further or extended visitors' or temporary workers' visas - or, apparently, any other kind - for the rest of our planned stay. We could, however, seek temporary citizenship.

The first step towards this was, via a quick detour to a photographer's studio, another urgent visit to police HQ. There, upon having signed umpteen bits of paper, and supplied five sets of beautifully-posed photographs, we - all three of us - also had to have our fingerprints taken. My British cultural mindset revolted as I protested vigorously about such an imposition, telling a succession of officers that there wasn't a shred of evidence to the effect that we - any of us, and never mind the rights and capabilities of minors, and all the rest of it - were, or ever had been, nor ever intended to be, criminals. But to the contrary, the chief of police himself proclaimed, with a logic which quite escaped me: this was for our protection. He also added that one copy would be sent to Interpol, for yet more reasons which were never quite made clear. I do wonder if our prints are on their database still.

By this time, fortunately, we'd pretty much got used to the Brasilian way of doing things; or, at least, to their inevitable accompanying annoyances, frustrations, and solutions. With our citizenship papers duly granted, we quickly shrugged off the whole encounter - an odd blessing in disguise, as it later turned out to be - and settled back into a routine far more relaxed than in our earlier days.

We'd recently acquired a television set - far from a given, back then - so were all able to enjoy those programmes which offered English subtitles. We also started going out to the cinema, not only availing ourselves of helpful subtitles there, too, but often, when an American film was showing, enjoying the luxurious position of being able both to listen and watch at the very same time. I particularly remember enjoying Love Story: in fact, in order not to embarrass one of our American friends, who'd proudly managed to get some apparently-rare tickets and invited us to come along with them, we actually went to see it twice.

And then, it was time for that great Brasilian festival: Carnival. Although the most famous show takes place in Rio de Janeiro, every city has its own parades and accompanying customs. In Recife, in the days leading up to the main event, we awoke one morning to the noise - nay, cacophony - of horns, squeakers, and whistles being blown in the roads all around our flat. Walking down to the beach, we then noticed that many of the cars travelling along the main road had suddenly had all of their doors removed. This, clearly, was to better aid their passengers, who were now leaning out at all sorts of precipitous, and definitely insecure and unsecured, angles whilst shouting and playing - in the loosest possible sense - various assorted instruments, and generally making as much noise as they possibly could; not to mention also enthusiastically waving flags, throwing streamers, and squirting water at anyone who happened to come within range. It was usually water, anyway; sometimes, needs obviously must, and it was instead beer, Coke, or a number of other unidentifiable things which one could only hope were hygienic enough. The soaked recipients of this latter popular pastime of the period would, however, normally be pre-armed themselves, often with substantial, solid, sizeable, pump-action squirters - although large, squeezy-type bottles would suffice at a pinch - and so, eagerly and gleefully, if rather indiscriminately, immediately shot back: and woe betide anyone caught in the crossfire. Unless, of course, they had a squirter, too. Otherwise, it was just yet another of the many gauntlets to be run at, on, or even in, any street, shop, school, or social occasion during the build-up to the celebrations proper.

On Carnival evening itself, I drove the three of us the short distance into the city centre, where we stood watching the colourful, and even more noisy procession. Absolutely everyone, except us, seemed to be blowing something, if not several somethings, all at the same time. The overall atmosphere, thankfully, was nothing but fun-filled and friendly: not at all the intimidating event which we'd been warned that it could sometimes be. And we didn't even get wet. We returned home in time to be able to see a good part of the main Rio procession on our TV: but it was, everyone later agreed, an experience far less exciting than witnessing our very own, if slightly smaller, event live on our own home ground.

The locals, of course, carried on partying and dancing in the streets and clubs until well into the next day. Pretty much, if not absolutely anything could and would go on; or come off, as might have been the case. It apparently wasn't at all unusual to find many men finishing up somewhere with someone else's wife. Then again, carnival does literally mean farewell to the flesh, and the next day was the start of Lent.

Not too long afterwards, in June, there was another great festival: São João (St. John, to us). This, it turns out, was also a day to be taken quite seriously, especially in the north-eastern regions. Luca invited us to share the 'São João lunch' at his house, over on the far side of the city from where we lived. What we hadn't realised in advance was that this was a meal regarded much like our own Christmas dinner: and it was most definitely as enjoyable an occasion. The local wine was plentiful, and the dishes traditional, all as far as possible being made from maize or manioc, yet amazing in their variety, from jellies to cakes, and everything in-between. The main meat dish was a broth made from chicken and vegetables. Its thick liquid, however, even considering the copious common black beans contained within, was an unexpected shade of deepest red. Apparently, Luca explained, when they used a fresh chicken, they made sure to use up everything, blood included. Eleanor, at this point, suddenly became rather less hungry. I don't know why: she quite happily ate black pudding in England.

Whilst driving over earlier, we'd seen a lot of people busily stacking up wood outside their houses. Most of these large piles consisted of strange, pre-constructed cubes of assorted loose timber, each about three foot square and bound with string; all especially stocked up by the local stores for the occasion, I later learned. We passed them again as we drove home, after dark. The huge, flaming mounds, it has to be said, were not only dangerously close to the buildings, but significantly constricting the already-narrow roads. Fireworks were also constantly being let off everywhere - we'd had nothing but constant firecrackers and little paper-wrapped explosive poppers in the streets for the past month, so by then we were past even taking notice - and the various social clubs, each holding their own special evening's entertainment, frequently sent up giant, blossoming rockets as well. It rather reminded us of Bonfire Night back home; at least, as we knew it then, when we all did our own thing in our gardens as a family, or maybe gathered with a few friends and neighbours, instead of attending the larger, organised parties which we've now come to know.

Talking of fires, we turned on our TV one morning, not all that long afterwards, to be confronted with the blackened remnants of a tall skyscraper block - it was still, in fact, burning - together with repeated, and at times quite distressing, news coverage of all that had happened during the preceding night. This was the first of two real, and very deadly, 'Towering Inferno' incidents which occurred in the city of São Paulo within the space of two years. Many survivors of this fire, I remember it being reported, especially those living on the upper levels, hadn't been able to use the stairwells to effect their escape. They were therefore forced to flee to the roof, from where, fortunately, they were later rescued by a helicopter.

I couldn't help but to think of Paul Quelt, our American banking acquaintance, and his company's sizeable fourteen-storey building over in central Recife. Only its bottom three floors were connected by a stairway, he'd bragged to me some months before. Anything above was served solely by lifts. This, he'd gone on to explain, in that very typical Brasilian way of the time, merely demonstrated how modern the establishment's owners - if not, I recalled, the local electricity grids - were. After all, he'd continued, people in this current day and age simply shouldn't be required to use such old-fashioned methods as exercising their legs. I often wonder if he'd eventually had emergency stairs and other measures installed, just in case they also ever suffered a fire; or, indeed, a power cut.

It was in the lead up to Carnival that we received another invitation from - or via - the British Club. This time, the pleasure of our company was requested on board a Royal Navy vessel which was soon to dock at Recife, and would be holding an onboard evening reception, prior to a formal 'sunset ceremony'.

The pleasure, we very quickly replied, would be all ours.

Things, from the ship's point of view, weren't quite so pleasing come the actual visit. Whilst ashore at the preceding stopover, Rio, one of their sailors had been shot and killed in some kind of targeted, anti-British attack. Consequentially, they had departed under severely-strained relations with the Brasilian authorities, and with his body still aboard. Feelings were obviously still running high. Matters weren't then improved in the slightest when the large vessel, a former aircraft carrier, reached Recife, and the 'port captain' (the harbour master, or some such equivalent) promptly refused permission for her to enter. The Brasilian navy's own carriers couldn't get in through the harbour entrance, he stated knowledgeably, and therefore this British one certainly wouldn't make it.

The opinion of HMS Triumph's captain, together with that of his entire crew, was that they could have brought it in sideways.

Nevertheless, at sea they remained, at anchor some good way out, with only their smaller - if still considerably-sized - escort ship allowed to come in to dock. It was therefore aboard her that we instead went for the reception, sipping drinks whilst the Royal Marines' band played on the quayside, and a few Brasilian naval personnel lounged on alleged guard against its buildings, their rifles slung at ease. The white-helmeted bandsmen marched to and fro as the sun sunk down, and the buglers blew out the appropriate tune as the ship's flags were lowered. It felt strangely emotional to stand alongside our own military countrymen, watching and hearing them playing our traditional music, as darkness fell over a foreign land which to us was also currently home.

The final act of the formalities was the playing of the national anthems: first the host, and then the visiting country. The Marines and crew, of course, as well as we civilians, all stood to attention for both. The Brasilian naval guards, however, seemed utterly confounded at our sudden strange change in behaviour come these two last pieces of music. Some continued to shamble about, some stopped and stood at ease, some shouldered arms, and one even went as far as to put his weapon down on the ground, stand rigidly, and attempt a salute. It appeared that most of them didn't even recognise their own national anthem, never mind know what to do when confronted with it.

The assembled ship-board company all concurred that, when it came to things like this, we Brits still did it best.

Knowing how long things could take to process through the Brasilian administrative system, I began preparations for our planned final departure a whole two months in advance. Several months previously, I'd received an urgent message asking me to go at once to Puerto Rico. It was a trip which all of my Brasilian contacts, even Luca, had firmly maintained to be impossible; the exit bureaucracy involved, certainly in terms of the required timeframe, was simply insurmountable. As it was, and never so easily beaten, I'd fallen lucky, finding a travel agent who took it as a matter of personal pride to overcome exactly such obstacles. True to his word, exit visa in hand, I was flying to Belem only two days later, onwards bound to San Juan.

I therefore engaged the same man's services again. Just in case.

The main papers required to effect our departure were the taxation forms, officially certifying that we'd met all of our dues: or, as applicable in our case, confirming that we'd never owed any. It was, as I've referred to before, a common and frequently-encountered practice - and occasional nightmare - wherever I went, no matter how short or long the stay. It also wasn't a wholly-unusual occurrence for me to have to explain how it all worked: but it would, I strongly suspected, take some considerable time for the average Brasilian civil servant's mind to grasp.

For some reason, not fully understood either then or since - except perhaps to say that, with a good dose of hindsight, my helpful 'fixer' did seem to know what he was doing - I ended up having to go and see the same police chief whom I'd previously encountered whilst obtaining our citizenship. He was, this time, extremely helpful, not at all concerned about being bothered by such an important and respected servant to his country as me - the man whose fingerprints he'd not so long ago sent off to Interpol - and heartily reassured me, in a manner which made me question if he and Luca weren't somehow related, that, as it was his department which put the final stamp on all papers granting anyone permission to leave the country on a permanent basis, there were, and would be, no possible problems at all. He also informed me that, just to make things even smoother for us, a new law had recently been enacted to the effect that wives who'd never held any employment in the country were no longer required to present any taxation clearances at all. And then, as if to put me even more at my ease, he gave me his private address and phone number, saying that if any future trouble should arise - which, of course, it shouldn't; but this is, after all, Brasil - he would immediately sort it out.

I, in the meantime, was busy rounding off my outstanding work, and settling in my recently-arrived replacement, Peter Hendrikes, trailing him around on as many introductory visits to our various potential and actual clients as I could possibly manage. Peter, a Dutch factory chemist, seemed very enthusiastic indeed about his new role: but we could already sense that his wife, Emma, wasn't so keen. As things turned out, we later heard that she'd left again after only a few months, saying that it simply wasn't like home. Which, of course, was true. Peter, though, remained in Brasil for many more years.

We were due to leave Brasil on a Sunday evening. A somewhat mournful Roberto transported us - along with our considerable amount of luggage, accumulated clothing, souvenirs, ornaments, stools, hangings, pots, assorted presents, and all other miscellaneous portable possessions - from our flat to the airport, where we'd also arranged to meet Luca for the very last time.

Not entirely surprisingly, there was no sign of my Brasilian colleague when we arrived. With nothing better to do, and taking advantage of the lull before the usual last-minute crowds, we checked-in for our flight, and then - and still no Luca - proceeded to the desk where our passports would be stamped for exit. The emigration officer took mine first, looked over the tax exemption certificate, and duly impressed his mark.

And with that one simple act, for all legal intents and purposes, I had left the country.

Well, not quite, and certainly not so simply, for he then turned to Eleanor's passport, and asked where her tax papers were. I repeated what I'd been told about the regulatory changes. This particular official, however, had obviously missed out on the memo. He just shook his head in response to my explanation, and stated that he now had no option but to refuse Eleanor and Raichelle - the latter also travelling on her mother's passport - leave to depart. What's more, he pointed out, although we hadn't yet passed the final emigration gate, I couldn't now be allowed to remain with my wife and child - all two steps away as they were - without going through the entire rigmarole of reapplying for my visa, citizenship, or whatever else I needed to be let back in to the country again.

I politely suggested that, then positively asked if, the officer could ring the police chief and easily clarify the matter. I even gave him the number. I would have thought, especially in Brasil, that a man, never mind a foreigner, in possession of the chief of police's private home number might just have caused a little raising of the odd eyebrow, along with the granting of a few respectful concessions. But, no: I just had to encounter the one awkward, obstinate, exception to the rule. He merely handed back our documents, and began to serve someone else.

Where was Luca when you needed him most? Right now, there was most definitely a 'problema', and I required urgent assistance. In desperation, therefore, with my only, if surely reasonable defence being that I most definitely had not yet left the country in any physical sense, I turned around and ran back outside the terminal before anyone could think to stop me.

And there, thankfully, was what I'd hoped to find: the faithful Roberto, still sitting in the car, apparently having been instructed by Luca not to leave until we were confirmed safely in the air. Thinking about this afterwards, I do wonder if, whatever my counterpart was up to at the time, and despite all of his assurances - never mind his doubtless genuine, earnest desire to bid his enthusiastic, Brasilian-style farewells to his esteemed friends - Luca already suspected that he wasn't likely to finish before our potentially still problem-ridden departure.

Dashing over to the car, I told Roberto to drive me as fast as he could through the usual chaotic traffic to the police chief's house. (And, no, with hindsight I don't know why I didn't try phoning first. Perhaps there simply wasn't an available working public telephone? Or perhaps I, now as a potential illegal immigrant, was just too busy running away?) To my huge relief, the chief was at home: in his slippers, watching TV. Having heard my problem, he stated that there was now only one thing to do, since, in his view, the airport's border control officers were just a load of idiots who still wouldn't understand the new arrangements even if he went with me in person to explain them.

"We'll go to my office," he announced, joining me in the car, slippers and all.

We drove at even greater speed - well, we did have the city's most senior policeman with us - to the main police building. We then had to stand outside and hammer on the door for several minutes until a sleepy-looking guard came and unlocked it. Once up in his office, the chief opened a desk drawer, extracted his official stamps, and proceeded to validate Eleanor's passport with full authority to vacate his country.

By now, of course, we were even further away from the airport, and we still had to return the police chief home. I'd already given up any hope of catching our flight, my thoughts instead now focused on the question of how, and how quickly, we - never mind our highly-excessive amount of baggage - got another. And where, in the meantime, could we sleep and eat? And how exactly did one secure a hotel room, or similar, when one didn't any longer, as far as Brasil was concerned, officially exist? Who else could, and should, I contact at this time of night, on a Sunday, to help us out?

Ellie and Rai stood waiting as our car once more drew up outside the terminal. But so, too, did another very welcome figure. Luca was so embarrassed and profuse in his apologies for not having arrived in time to prevent my hurried unofficial exodus; not to have been able to save me, and us, all that trouble and worry; not to have been there just one final time to present his famous badge and card, to circumvent all the usual trials of officialdom, and see us safely through to the plane.

"But don't worry," he proclaimed. "There's no problem!"

For once, I really hoped that those famous words would pretty much be his last.

"Take your time," he assured us. "There's no rush. I have already given instructions that the plane must await your return, however long it was. They will not take off without you. And you must be taken in First Class, to make up for all the trouble that you have had because of our terrible Brasilian bureaucracy."

He then led us, with one more cheery "Come on, Doutor Franco!", straight past security - without anyone having to undergo any form of luggage or personal search or other hindrance whatsoever - straight through the departure gate, and straight out onto to the tarmac; right to the bottom of the aircraft's steps. There, and only there, with the crew patiently watching on from the doorway above, did we bid our long goodbyes.

Upon our eventual boarding, we were ceremoniously ushered all the way to the front by a suitably-fawning stewardess, who ensured that we were all properly and comfortably seated, settled, and appropriately refreshed, before telling the pilot that all was now finally well. The engines were immediately started, the jet soon taxied out, and it wasn't long before we were looking down at the lights of Recife, fading out below us for the very last time.

I almost expected to be able to see Luca and Roberto still waving.

It wasn't actually my last ever trip to Brasil. ISSCT's three-yearly conferences were always staged in a different sugar-producing country, each organised by its host. 1977's meeting was held in São Paulo. Being the sixteenth such event, the techniques for putting together all of its various components should surely, by now, have been pretty much perfected. Only this one, of course, was subjected to Brasilian influence. It was, fundamentally, utter chaos. Firstly, the programme of presented papers was such that it was impossible to attend some without missing others. Then, although the conference's standard language is English, many of the papers were presented only in Portuguese, and there weren't even any printouts available to take home and have translated later. Internal travel, accommodation, and feeding arrangements proved to be more than hectic - certainly not to the efficient and impressive standards of my very first Taipei experience, never mind the others which I'd also previously attended - and, altogether, the delegates' tempers just grew worse and worse as the week wore on. I had hoped to revisit my old stomping grounds up in the north east, but, in the end, that also proved too difficult to arrange, so I flew back out from São Paulo airport, never to return.

Although the country has undoubtedly changed so much, and in so many aspects, I'd still like to go back once more, if only to see how much Recife and its environs has been developed over the years. I now see it regularly featured in magazines and holiday brochures: and developed and changed it certainly has. Old Boa Viagem square is still there, albeit smothered by surrounding swathes of tall, sky-scraping apartment blocks, stretching as far down the coast as can be seen. The quiet residential area where we once lived is now reported to be teeming with cosmopolitan city life in all of its various modern glories, complete with a vast shopping mall. Our dark, once-empty beach is apparently constantly-illuminated at night, and its calm, warm, sheltered waters are, by day, now renowned for shark attacks, and often unsafe for swimming. I suspect that those old idyllic, deserted, palm-fringed beaches are not what they once were, either.

It would certainly be an interesting trip.

13

WEST INDIAN WORRIES

Out of all my visits and travels, the Caribbean - or, more specifically, its islands - was the region which I ultimately felt that I'd best come to know. It was also, barring that initial Mauritius trip, where I both started and truly ended my overseas career: my first proper commercial visit, and my last ever consultancy trip. In-between, even whilst in Brasil, I journeyed to the 'Windies' at least once every single year. The main and simple reason was that Tate & Lyle and their various group companies owned, or later managed, a good many sugar estates throughout Jamaica and Trinidad. In time, they and I formed several additional connections amongst the other islands, which in due course extended out into their various governmental departments, who in turn also then wanted studies and surveys for themselves.

For some reason - probably just because it sounded good - I named this last chapter 'worries'. Yet, truth be told, and despite the archetypal lack of such states of mind within the Caribbean cultures anyway, I actually encountered very few problems during my frequent presence in this part of the world. With one notable exception, in fact, nothing too much untoward or adventurous ever really happened at all. Perhaps this was simply because I always stayed with, and was very well catered for and entertained by, members of the local staff, and thus managed to avoid any of the more notorious - and also, with some disappointment, the more glamorous and well-known - parts. Still, as with most of the other countries which I visited, that also meant getting to know them and their peoples as they truly were, not simply for their sunshine, palm trees, swimming pools, beaches, rum punches, and tropical oceans; their hotel and bar staff, shop-keepers, and tour guides.

It occurs to me only now, having just written that last paragraph, that one of the things which I've never really touched upon before, but certainly still bearing mention at some point, is the rather obvious question of 'colour'. This, in the main, if maybe surprising to some, is simply because one's race or ethnicity, never mind the actual colour of one's skin, just never seemed to enter into personal or business relations as far as I ever encountered them; not in the West Indies, and pretty much not anywhere else. I stayed with so many different Afro-Carib members of our companies' and sites' staff over the years, often becoming good and close friends, and I always felt totally at home, being made easily, genuinely, and exceptionally welcome not only in their houses, but by their neighbours and relations, too. More generally, no matter who people were, or of what position or background or appearance, I was always treated as nothing but an equal, no more and no less; no different, as far as I was ever aware, in any way at all.

I did, however, learn a salutary lesson during my very first Jamaican trip, when one of the local field managers, a large and very dark-skinned man, was telling me of his troubles with the estate's labour force.

"These lazy black n*****s are the problem," he pronounced, sourly. "They're not like us white men, you see."

I looked up at his face in amazement for a moment, before realising that, to him and to his colleagues, 'white' was not necessarily a matter of pigmentation, but of education and position in society.

Make of that now what you will.

Trinidad, unfortunately, did have a 'coloured' problem: but it wasn't necessarily what one might initially have expected, either. Nearly all of the European, Creole, mixed race, and other more-or-less indigenous 'white' population - themselves probably descended from ancient Central to Southern American immigrants - were to be found in the management sectors of industry and commerce, where there was also a good proportion of those with Chinese roots. The majority of the rest of the population, and therefore employment sectors, was split approximately equally between peoples of Eastern Indian and African ethnicities. These latter two groups lived mostly in completely different villages and townships from each other, their separation not merely social, but extending well into the political field. The Indians stemmed from indentured labour brought into the island after the abolition of slavery, whilst the Afro group, unsurprisingly, was derived from the slaves themselves, arriving in the late 18th century to work on the cotton and sugar plantations. This historical difference left a distinct and overtly-manifested feeling between them - not necessarily shared by everyone else - that the 'slaves' were still of a lower class than those of their Indian replacements; and, from the Africans' point of view, that they were still being treated as such.

It was bound, one day, to come to a head.

The extent of T&L's Trinidadian estates was considerable, stretching many miles both to the north and south of our Brechin Castle factory, itself located halfway down the country's western coast. The entirety of those holdings - the whole of the island's sugar operations, at the time - was under the control of a company called Caroni, and, early in March 1970, Patrick Insh was its agricultural manager.

It was Pat who, as usual, met me at Piarco airport, about twenty miles east of the capital, Port of Spain. Straight from the start, things felt far from right. He wasn't at all in his normal conversational mood, but seemed oddly and obviously wary, not to say increasingly nervous, especially when we drove through the several Indian-populated towns and villages along our southerly route. And then, on finally reaching his family home, my usual accommodation when out on visits, he reached into the car's glove pocket and took out a hand-gun: something which I'd never associated with him, nor any of my overseas colleagues, before.

Only once safely inside his house did he explain that there was known to be trouble brewing between the local Indians and 'Blacks'. More to the point, it was looking increasingly like the sugar company's management might somehow end up getting caught between them. Nationally, issues had been mounting over the last couple of years; more so as this particular one had begun. A general sense of discontent was now accompanied by numerous demonstrations and protests all over the island. A 'black power' movement was gaining momentum, about the same time as it was starting over in the USA and elsewhere, and there were all kinds of other both separate and interacting socio-economic and political elements, concerns and motivations: disaffected youths and workers, urban poverty, increasing trades union militancy, demands for 'people power' and general societal change, cries to stamp out racism and inequalities, challenges to the very long-standing government and its one and only prime minister since the country's 1962 independence, general ongoing post-colonial gripes... And all the usual, quite possibly justifiable, rest of it.

About a week later, it was announced on Island Radio that the Blacks were planning to hold a mass rally, assembling early one morning in the capital's Queen's Park Square. Come that actual day, and the radio news further confirmed these protesters' apparent intentions to march on Caroni's factory, the site not only of our main company office, but adjacent to a large staff housing compound. Their purported end objective, it seemed clear to us, was nothing short of a full-scale take-over, by force if need be.

As a manager, Patrick preferred to keep a personal eye on things, rather than just sitting in his office and dealing with his staff by radio and telephone. The total area under his care was, however, far from easy to cover from the road. And so, like many others in a similar position, he'd resorted to the most practical solution: first, he'd learned to fly, then he'd hired a two-seater, high-wing Piper from the aerial fire-fighting and crop-spraying outfit who were contracted to Caroni, and based out of one of the estate's airstrips. In fact, I often went up with him on his early morning rounds over the cane fields, inspecting the ground conditions, monitoring the activities and progress of various work gangs, and speaking with their supervisors on the radio. On this particular morning, though, the cane fields weren't at all uppermost on his mind. Not long after hearing that radio broadcast, he'd received a call from the island's police. Would he be able to take his plane and fly up to and over the island's north-western tip, they'd asked; to the areas around Port of Spain, in other words. Apparently, because of the vast crowds, the police couldn't get through to our side with their vehicles, and they needed someone to report on the current situation, then to continue monitoring ongoing progress. Presumably, they either also couldn't get to, or simply didn't have access to, a spare aircraft of their own.

Pat asked me if I'd accompany him, so we both took off from the factory airstrip, and were soon passing at a fairly low altitude over Queen's Park itself. It wasn't hard to confirm that the hoards of people now streaming out of it, on foot and in cars, carts, and all kinds of other various means of transport, did indeed seem to be heading straight in the direction of Brechin Castle. Port of Spain was about twenty five miles away, an hour or so's drive by car. The main crowd, we estimated, if they were determined enough, would reach us by mid-afternoon.

We flew back as quickly as we could, and I, albeit with deep reservations, then set off in my Land Rover to carry out a pre-planned tour of the estate's southern parts. After all, there was nothing else to do, and it at least served as a kind of distraction activity; for a while. Pat, meantime, stood ready to carry out further reconnaissance missions, as and when required.

Shortly before midday, the estate's general manager received a telephone call from the leader of a predominantly Indian trade union, a man reputed to be involved in various dubious and underhand, if not outright criminal activities; and a distinct thorn in the sugar company's flesh for many, many years. He informed the

GM in dramatic, and very definitely non-optional terms, that he would shortly be arriving at Brechin with some of his men, in order to "protect Caroni".

How terribly thoughtful and kind.

Or, to quote more directly: "Let any of those Black b******s set one foot in your place, and they'll wish that they'd never thought to come."

Not long afterwards, a fleet of large American cars drew up in front of the main office, immediately disgorging a considerable number of Indians. Proceeding just as quickly to open up their vehicles' boots (or trunks, if you prefer), they withdrew an equally-great number of entirely-illegal firearms, including fully-automatic rifles. And thus armed, his henchmen then stood guard right outside the office doors all afternoon.

By now, I'd returned from my occupational therapy, and was sitting in said offices, along with a good many others, simply waiting and worrying. News of the Blacks' mass procession was received through various sources from time to time. Thankfully, it seemed to be slowing and thinning out, the distance of travel and growing heat of the day no doubt causing many to become tired, hungry, and generally discouraged. Late in the afternoon, however, we heard that its advance party had finally reached the town of Couva, just a short distance to the north of us.

And there it apparently stopped.

A little while later, a message from the protesters was relayed up to us, asking if representatives could be sent to meet with our management. We nervously agreed - well, at least they'd asked first - at the same time wondering what fate awaited us, what options would be presented to us, what chances we had.

Arriving in due course, and thankfully passing our armed guards without incident, the men - a small, proud, determined, but nevertheless quite polite group - cleared their throats and stated their demands: to the basic effect that they wished to register a complaint about the number of their representatives in the factory and field departments being far outweighed by the number of Indians there.

Oh. Right. Duly noted. Thank you. We'll see what can be done as soon as possible.

And that, it seemed, was that.

I must at this point add a form of caveat; at least, an explanatory note. The above account is how I remember it, how I and we perceived it - if maybe from a simplistic point of view - as it happened, whilst I was actually there. However, on checking a few details recently upon that wonderful technological development, the internet, I find that recorded history seems to recall events rather differently in many respects. Yes, this was definitely the 'March to Caroni' of 12th March 1970: but exactly which Caroni the protesters had planned as their original destination now seems rather more open to question, there being the choices of Caroni the sugar company, headquartered at Brechin Castle, Caroni the general administrative and sugar-growing area, or Caroni the small Indian-occupied settlement situated several miles east of the capital, out towards the airport. It is generally now stated - which accords with the apparent nature of its intentions - that the march was heading for the latter; but then, upon having reached it, they decided to carry on, turning back west and south to pass through the Indian towns of first Chaguanas and then Couva, a stated gruelling distance of some twenty six miles in total. But no overt mention of the nearby Brechin Castle factory, nor the Caroni company, ever seems to be made.

Questionable, too, is the protesters' true purpose. It appears now that, far from being purely a Blacks' or Africans' protest, an attempt solely by them to seize more power and influence, or even just to protest against a government which they disliked (although, apparently, not as much as the Indians did), its aim was more to serve as a rallying call to national unity, in particular amongst the two main racial classes; a call for equal rights; a show of solidarity between the two 'underclasses', some say. How much the Indians supported this at the time is unclear, although it is reported that one particular Hindu activist and MP, stemming from Caroni village, and also the then head of the Trinidadian sugar and factory workers' trade union, was very much against the march, and thus rather keen, along with others, to incite violence within it. (And also seemingly keen to protect the interests and incomes of his sugar industry members in a more direct sense, as we rather personally appear to have found out.) Happily, this feared outcome never transpired, and with specific pleas by the protesters' leaders to respect the Indians, the 40,000 reported attendees apparently all proceeded in a most peaceful and disciplined manner.

The March was also, some say, a call to throw off the remaining colonial shackles which continued to preserve such divisions and inequalities within the island's society: such influences obviously including T&L's sugar estates. Indeed, later in the evening, long after our unplanned visitors had departed, in some park or square at Couva, so it is written, 25,000 or more of the marchers held a 'People's Parliament', where it was resolved that their sugar industry's British ownership must end. No-one, to my knowledge, ever came and said that to us in person, although it does certainly tally with what we'd been warned.

I'm sure that the real politics and motivations were much more complex than I've made them out - than even the few articles which I found and read have made them out - and history is always told with hindsight, and often recorded by those with particular viewpoints. It's also often recorded as simple black and white facts, whilst, on the ground, reality may well have been a much more muddled shade of grey. As to which particular parts of both my and others' accounts are open to dispute, debate or interpretation, therefore, I cannot say more; nor which, if not all, are factually correct.

All that I can say now is that for us, the management, staff, and workers at the Caroni sugar factory, matters quickly settled back into their usual state of affairs, and no more major issues occurred: not even during the soon-following 'Black Revolution', the island's general uprising and subsequent State of Emergency, the many further tensions, issues, and changes which were to come for the country both immediately and farther down the line.

But the People's Parliament did ultimately get its wish. In 1975, T&L sold their entire Trinidadian holdings back to the country's government. And sadly, for a good number of reasons - as happened throughout nearly all of the Caribbean and West Indian lands - these long-standing sugar-producing activities, and all that went with them, for good or ill, slowly began to decrease over the following decades.

The entire Trinidadian industry finally came to a complete halt in the mid 2000s.

Back on much more mundane, day to day matters, Pat, as you might expect, sometimes used to fly me up to the main airport when I was on my way back to the UK. It was towards the end of one of my later visits that he began having trouble with his plane's starter motor. Unable to obtain the necessary spare part, he was therefore forced to get someone to swing the prop every time that he wanted to fire up the engine. It wasn't a particularly onerous operation, but it wasn't always easy to find a spare volunteer for the job, either. Consequently, he tried not to switch off the engine unless he knew that he'd finished flying for the day, content to let it idle whilst quickly holding a conversation, or unloading his cargo, documents, or passenger - in this case me - before taking straight off again.

Which is also exactly what he did later that same day, so I'm told, upon arriving back at his own airstrip and passing on the plane - engine still ticking over - to another young man who occasionally used it. Pat also mentioned that the fuel tank needed refilling: which the new pilot promptly arranged. The mechanic responsible for the task duly dragged the fuel hose over from his tanker; and, of course, straight into the path of the still-slowly-rotating propeller. This immediately sheared the hose straight in two, showering the engine cowling with aviation spirit, which, in split seconds, burst into a ball of flame. The plane, and the entire fuel station, were apparently destroyed in less time than it takes to tell.

Not surprisingly, this was the end of Patrick's flying career, as

Caroni decided neither to replace the plane, nor to continue the contract for the 'flying fire brigade'. Not long afterwards, and completely unrelated to this mildly-embarrassing incident, Pat himself retired.

It was therefore his replacement, Geoffrey Fraser, who, some years later, had to come to my rescue when the government suddenly decided to get tough with the number of 'non-Trinies' trying to enter their island. My visit had been timed to coincide with that of Ian Whiter, one of our factory engineering consultants. I, however, was flying in directly from the UK, landing well into the evening, whilst Ian's arrival, via the USA, was scheduled for early afternoon.

The immigration procedures at Piarco were, as usual, lengthy: queues of anything up to three quarters of an hour were frequently encountered before one finally reached the desk. When it eventually came to my turn, the immigration officer, a not-too-highly-placed policeman, enquired, as was their wont, why I wished to visit his country. I told him that I was carrying out my normal two-week consultancy trip.

He asked to see my work permit.

I tried to explain that I wasn't there to 'work', in terms of what would necessitate such a permit, and that I wasn't being paid in Trinidad, nor working for anyone based in Trinidad, nor harbouring any plans to carry out any other activities, nor to take up any future engagements or residence or anything else in Trinidad, and all the rest of it, and that we had already long ago cleared this with the authorities, anyway...

But it was all in vain. After a long and increasingly-frustrating harangue - which, oddly, never even once mentioned the word 'taxes' - I was taken off into a separate office, and there, with a different but more senior officer, as the night-time hours drew on, and the airport became more and more devoid not only of fellow passengers, but of any other people at all, went through the entire same series of questions and answers all over again.

And again, it was all to no avail. In spite of all my protestations,

I was told that my presence was no longer considered desirable in the country, that my passport would accordingly be impounded, and that I would be put onto the next plane back to the UK the following day. And in the meantime, lest I tried to abscond, I would be detained under guard at the airport's Belle Aire hotel.

At least this last part didn't present a problem: I'd already booked a night's room there in any case, thinking to save someone having to come up to meet me so late.

With no more options remaining, I asked only that the officer telephoned Geoff down at Brechin to confirm my story; and, more importantly, to inform him that I was being held in custody. Not that it'd get done until the morning, of course: but then again, that wasn't all that many hours away. I also mentioned that another colleague would already have come through earlier in the day. Had he been allowed entry?

"You'll find him in the hotel, too," came the terse reply.

And indeed I did: absolutely fuming. Ian was rather deaf, and his disability often made him more than a little irascible. The inevitable outcome obviously hadn't further enamoured him to the authorities one bit. Having exchanged our experiences, and consumed several well-needed drinks to calm our over-frayed tempers, Ian added that he had at least been able to telephone through to Caroni's offices, and that, although nothing could have been done right then, so near to the close of the working day, they had promised to contact the necessary government department first thing on the next.

Something or someone clearly did the trick. Having been periodically woken and checked up on by a scowling policeman all through the remaining early hours, we were just finishing up our rather late 'prison' breakfasts when yet another immigration officer arrived, with Geoff in tow. This particular official almost - almost - apologised for his previous colleagues' incompetence. By way of an excuse, however, he pointed out that it was Caroni who had utterly failed to inform the proper persons - whoever, most unhelpfully not stipulated, they might actually be - by way of the proper processes - if only anyone had ever heard of them before, either - of our impending arrival. But Geoff kept his mouth firmly shut, and Ian and I we were handed back our suitably-stamped passports, finally free to commence our duties.

They never tried that one on us again.

****

They didn't want to let me into Barbados, either. It wasn't by far my first visit, I knew the required drills, and I'd never encountered any problems there before. Presenting myself at the airport's immigration desks, therefore, I wasn't surprised to be met with the usual enquiries as to where I planned to stay. For the standard foreigner entering the island, whether on business or for leisure, this question really meant 'at what hotel?' If one hadn't already been prearranged, then the new arrival was immediately directed towards the long row of directly-linked telephones which lined an entire wall of the immigration hall, from which a booking could easily be made. It didn't seem to matter where, nor for how long, as long as an initial reservation existed somewhere.

I, however, had to inform the officer that not only didn't I have a hotel reservation, but that I didn't wish to make one, either. I'd already made plans to stay with Dr Callum Hicks, from the island's Sugar Research Station.

The immigration officer demanded proof.

"Well," I responded, in the absence of anything else more formal, "He'll be waiting outside to pick me up."

So the officer asked me to look out through a window - it was a very small and simple airport - and point to the man in question, then beckon him over to verify my claim. Which wasn't an unreasonable demand, I thought, and should surely settle the matter very easily indeed.

Except that Cal wasn't anywhere to be seen.

After what seemed an eternity, during which I could do nothing but to state repeatedly that my contact must just, obviously, be running a little late, and only moments before I was, yet again, about to be incarcerated and then packed off on the very next plane home, Callum's car finally came screeching into the forecourt in a cloud of dust. I waved him over with considerable relief, and, with but a few more short questions remaining, I was accorded admittance.

Cal, it transpired, had only just been released from police custody himself, having been apprehended at his house earlier that morning, and then most unceremoniously carted away to a cell at the main station, for the heinous offence of conveying passengers in an unauthorised public transport vehicle. It seems that he'd been seen and reported on the previous day, indeed carrying a number of persons in the back of his pick-up truck upon a public road: but for the sole purpose, as he was fortunately and quickly able to explain, of giving a wholly-non-remunerated, free-of-charge, goodness-of-his-heart-motivated lift home to some of his workers at the end of a particularly long and arduous shift. The police had begrudgingly returned him to his house just in time for him to be able to grab his car and dash over to the airport to meet me. No speeding tickets seem to have been involved.

Dr Hicks was a rather well-known character in any case, his somewhat eccentric behavioural tendencies from time to time surely mystifying both the police and several other local officials alike, yet clearly endearing him to a multitude of friends and colleagues. I remember once sitting outside his house with several of them, passing a good proportion of a very hot tropical evening singing through the entirety of Handel's Messiah, in all four vocal parts, in occasional harmony, accompanied by Callum on the piano, and others on the steel pans. As, of course, you do. Or, at least, as in those days we did. Please don't ask me why.

****

Puerto Rico was a country which I didn't officially visit much, but again frequently used for overnight stops en route to others. Apart from its excellent aforementioned rum distillery, it also had a rather high-class hotel, known as the Tennis Club. There, for the price of your chosen drink, one could often watch world-class players practising, not just during the day, but upon their floodlit courts in the evenings. I even got to stay there once, along with a colleague but, yet again, courtesy of Pan Am, without my luggage.

We'd both been summoned urgently - him from London, me from Brasil - to advise one of the local sugar factories on problems stemming from, of all things, a labour dispute. The workers had, for some reason, refused to start hand-cutting at the start of their cropping season, so the management were now employing all sorts of extreme measures to get the ripe cane off the land and into the factory before it went past its best. Ground conditions weren't at all suitable for mechanical cut-load harvesters, so they'd resorted to using bulldozers to firstly push down the stiff, tall stalks, and then grab-load them - simply ripping the canes up out of the ground, complete with dirt, stones, and all - into whatever transports they could physically get anywhere near. All kinds of damage were quickly accumulating, pretty much everywhere and to everything, the fields included. Solutions were desperately needed.

We were able to offer some suggestions for future developments in their cultivation systems, eventually making them suitable for mechanical harvesting, and also advised on cane cleaning procedures within the mill. But, although I believe that these were subsequently adopted, we couldn't do much to rectify their immediate issues, nor to help with whatever had caused them.

That season, so I understand, marked the start of the decline of the cane sugar industry in Puerto Rico. And so, apart from changing several more planes at San Juan airport, I never went there again.

****

My final visits to the Windies - to anywhere, in fact, as a sugar industry or general agricultural engineer - were, in many ways, my most enjoyable. They were certainly the easiest; and definitely the most leisurely and comfortable. But then again, they were also unpaid.

It was now the late 1980s and early '90s, and I had, so I'd thought, officially retired. Yet I'd hardly had time to get used to being permanently back home, finally free to put my feet up, when the volunteer organisation, British Executive Services Overseas, called.

Would I be prepared to spend a month on the island of St. Kitts?

It was a new destination for me, the trip both pleasant and interesting, and the work - advising on a variety of cane harvesting and transport matters - generally not too strenuous. I wasn't long back home, and they wanted me out for another two months. The following year, and the one after that, I was asked if I'd go out again on follow-up visits.

But at least, at last, I could take Ellie with me. We were given sole use of a very nice house near Basseterre, where, I must admit - the madcap terrors of speeding minibus taxis on their curvaceous 'countryside' roads notwithstanding - we both had a wonderful time. Not being a lady much given to lazing around in the tropical sun, however, nor, despite everything, still ever having wanted to learn to drive, never mind swim, Ellie initially found herself rather bored. But she soon found herself a new, wholly-unexpected, totally unqualified, but nevertheless apparently extremely successful occupation whilst I toiled away: she became a teacher at the estate's primary school. It was a position from which, we were told at our last departure, she would be sorely missed. In the meantime, amongst many other things, she earned the distinction of producing the first and only daffodils ever seen in that small Caribbean nation. True, they were made of wire and coloured paper: but, from an appropriate distance, they still looked very real, and certainly served to illustrate the Wordsworth poem which the local children were learning.

I spent one more two-month period out there, once more on my own. Staying mostly with Marcus Jeday, a former colleague who was now the estate's Chief Executive, I drove myself around in an old ex-British, ex-army jeep that had finally, I was delighted to discover, been fitted with the radio systems which I'd originally advised were needed over three years ago. Sadly, Marc's wife died shortly afterwards. Combined with the devastation sustained by the island's sugar industry during a particularly severe hurricane season, it was enough to make him call it a day and return back to England.

By the end of July 2005, the sugar industry on St. Kitts had also called it a day, and closed its doors for the final time.

Time was being called for me, too. Although invited by BESO and others to carry out much more consultancy work in this and other countries, I semi-regretfully declined. Whilst technically perfectly fit to travel, I'd recently suffered a very mild kind of stroke, after which I'd agreed to take part in a clinical drug trial, in turn requiring me to undergo frequent medical checks.

So thus it all ended. I would miss the travel, I knew: but those last trips had been not only extremely satisfactory, but also wholly uneventful, producing nothing at all noteworthy from my side to report.

And that, I thought, upon only very brief reflection, wasn't really such a bad thing at all.

POSTSCRIPT - A FEW FINAL FLINGS

Sadly, although he retained a great fondness for many of them, and would have loved to have returned just for a holiday, Frank never made it back to any of the countries in which he'd worked. He did, however, keep in close and lifelong touch with the very many friends and former colleagues whom he'd made and met throughout his career, often travelling to stay with them in the UK, and otherwise running up admirable phone bills, writing a good many long letters, and annually working his way down an extensive Christmas card list.

His retirement, the BESO work notwithstanding, was a far from inactive time. As well as his lifelong singing and church interests, Frank also discovered rather late talents for acrylic painting - several masterpieces now hang in various family homes - and amateur dramatics, both on stage and behind the scenes (where Eleanor, mostly only with feigned reluctance, also became a renowned 'mistress of the robes'). Writing, directing, producing, and making most of the props and scenery, it seems that he'd always held a secret hankering for the thrill of the greasepaint. He claimed to favour character parts, but the legendary dancing pantomime camel (rear end thereof) in his self-penned version of 'Ali Baba and the Naughty Thieves' was certainly a sight to see; as was another production's wonderfully camped-up 'Dame'.

Ellie, in all honesty, was never a great one for 'exotic' travel: those strange, foreign foods most definitely weren't to her taste, and she was even less a fan of flying. She seemed to enjoy the St. Kitts trips, and there was another long trip to Miami, too, where they stayed with Pete and Paula Wade: but that was enough for her. Compromise trips did, however, see the pair undertaking many a European sea or river cruise, often allied with further land stays and tours.

Although already well into his 80s, Frank continued to travel extensively - in fact, even more so - after Eleanor's death, not only throughout the UK, staying with his widespread friends and family, but undertaking many European city stays, embarking upon several more river and sea cruises, and then venturing out into the realms of long-distance train journeys. He completed several passages across Alaska and Canada (including shooting whitewater rapids on an inflatable raft), and then explored the Baltics and several former Soviet states. Still sprightly, he seemed content to undertake the majority of these trips all on his own, and found no difficulty in finding company, nor in forming yet more lasting friendships along the way.

Once, having departed a Russian cruise ship for a walking tour of Saint Petersburg, he fell and cut his arm. Having located the appropriate medical assistance, he was in the process of having his injury stitched up when the Russian doctor observed how the word 'wound' (as in a cut) was spelt just the same as, but sounded utterly different to, what you do when tidying a wire or hose or piece of thread around a reel. There apparently followed a long and apparently-fascinating discussion about the oddities of English language pronunciations, which he'd doubtless have loved to recount here in much greater detail.

He also, finally, if just a little late, got around to having a flying lesson.

Returning from a trip around Croatia, aged 90, he began to feel "a little unwell". Bar childhood pneumonia, and a minor TIA suffered in his home garden several years earlier - of course, given that not all of him felt 'funny', and as he could still move well enough, he insisted on driving himself to the doctors to get checked out - this was his first and only major health problem. Sadly, however, after a year and a half of slowly-declining health, several reasonably-major surgeries, and a final few days bed-ridden at Faye's countryside home, it was also his last.

Frank died in late 2011. His coffin was topped with a tribute of tropical flowers.

****

Amongst the many writings found when clearing his house was a letter which he'd written to one of his grandchildren, a typical formal response to some passing innocent teenage enquiry about his life and general beliefs. It seems pertinent to end with some extracts, as regards what his foreign work and travels had taught him.

"I learned so much from my overseas visits.

Firstly, that although one could be polite and respectful, there was never any point in being deferential. Also, that frankness, together with reasonably true and straight talking, was always respected: but any attempts at putting on pressure was a sure way not to get results.

I often had to work in close contact with many governmental officials, including senior ministers, and once a vice-president: in that particular case, a title which soon proved very apt indeed. One thing that I quickly discovered was that there was nothing to be afraid of in talking with such highly-placed politicians. It was the lower-placed officials who were always the most difficult to deal with.

I soon learned to see through falseness in those whom I met; and, sometimes surprisingly, this was again more often encountered with those much less highly-placed in the general scale of things.

I've come across some very shady characters in my time, not to mention many a 'bent' politician or manager. In fact, many with whom I dealt, including many whom I got to know quite well, were, to say the least, somewhat corrupt in their business dealings. But that was also often so very normal: and I knew it, and they knew it, and they knew that I knew it, so we all still got along very well - in fact, I came to like some of them very much indeed - and I just kept my disapproval, or usually mere wry smiles and shakes of the head, to myself.

But then again, "Let him that is without sin..." For many's the time when I've looked at my 'enlightened' home country from afar; and many, many times I didn't much like what I saw. I've so often said to people that you can't judge your own country until you get outside of it, and try to look in through others' eyes.

I met so many people over the years, of so many different nationalities, creeds, colours and cultures, and from them gained a good insight into their lives and thoughts. I often found myself listening to their problems and worries when they wished and chose to confide in me. They talked about the troubles which they met and endured within their own lives, and of those being suffered by their country as a whole. I always respected these views. But whatever the conversation, I also always had to bear in mind that I was only a visiting guest of the country concerned, and mustn't ever criticise it or its regime. Not whilst there, at least; nor if likely to be going back.

I quickly learnt that many peoples' cultures, never mind their strange ways and foods, are so often simply enforced upon them by their geography and climate, together with many other factors which are utterly and completely foreign to us. And then again, it always paid to remember that, to them, it was actually us who were the strange, foreign ones.

And whatever those cultures, whatever their ways, never mind their specific or wider problems, I soon realised, there is never going to be any single, simple, one-size-fits-all solution which can be so easily applied to very differing places.

More than anything, though, I concluded that most people, from all countries, of all faiths and beliefs, of all backgrounds, colours, creeds, and political persuasions, are by nature kindly disposed. Mostly. No matter where I was, or why, and no matter who or what they were, I always received tremendous kindness, hospitality, and friendliness, and generally found a great willingness to help, no matter what I, or we, had to contend with. Some of the people whom I most admired were, in some way or another, very different to me:

yet it never should have, and usually never did, made a difference at all. The fact that they may have thought, acted, or merely eaten differently from us is no reason not to accept them for what they are: human beings who, in spite of everything, must be respected just the same.

Tolerance, I learnt, is the key word.

And you never know: you might even get to like the food."

A Note From Rai, 2025

We knew that Dad was writing his memoirs. He'd started with some hand-written notes, made whilst on his last St. Kitts trip. One of my sisters then passed on to him some kind of old typewriter-cum-word-processor thing, which, eventually, albeit only after a good many years' use, he finally upgraded to a very basic computer, complete with its 'modern' smaller and not-so-floppy discs. He set them up on an old kitchen table in the smallest bedroom of his house, along with a variety of inherited printers, some sturdy, hand-built bookshelves, and a multiplicity of cardboard box files.

Proudly, one by one, at some appropriate family occasion or other, several large brown envelopes of printed, albeit often still hand-annotated manuscripts would be presented: *'Myself When Young'*, *'Into The Country'*, *'How We Won The War'*, *'Eleanor's Tale'*, and then *'Sweet and Sour'* (which subsequently, although very apt, I re-titled because there's just far too many of them about already). Being a latecomer to the family, my sisters knew and remembered much more than I about the names, persons, and places of his younger days, and so, except for keeping my copies safe, and with many time-consuming concerns of my own, I must admit that I never really paid them as much attention as I ought whilst I had the chance: for which I can only now express my deepest regret.

Still, as regards quite a lot of this particular work, I was the one who had been closest, still living at home for the majority of the times in question - not to mention being intimately involved on a few occasions - so it was naturally assumed that I knew most about those goings on.

That, it transpires, was a very bad assumption.

Anyhows, fast forward to after his death, and the obligatory house clearing. As is the way of these things, there was little time to look through all of the interesting-looking boxes which filled up his study's bookshelves. It was only several years later, when I was over at Faye's house, that I was duly informed of the urgent need to help her to dispense with "a load of Dad's old stuff" which still languished there.

And there it all was: slides, photographs, original technical drawings, plant and equipment designs, ancient glass printing plates, lecture notes, articles, reports, letters, mementoes, souvenir programmes, matches and cocktail sticks from just about every hotel and airline in the world, his own father's college and later field notebooks, ancient engineering and metallurgical tables, drawing instruments... Even his trusty old expanding leather briefcase, complete with many an airline sticker. Dad never threw anything away. This fascinating collection, once sorted and roughly catalogued, filled a very, very large, long, and extremely heavy cardboard box, which sadly I had no room to store. So I tried to find a good home for it, eventually managing to get his former employers to take it into their London-based museum-cum-store, where I hope that it still remains.

Amongst all of those papers there was another old manuscript copy of 'Sweet and Sour'. Interest piqued, I gave it another quick read-through, and was soon promising myself - as you do - that I must one day "do something with it". I'd already acquired a flash drive upon which some younger, and therefore cleverer-than-me relative had managed to save his old word-processor files, at least thrice converted from one out-of-date software format to another. It was still a bit of a task, but I eventually managed to get everything into a workable form, if sometimes line by blinking line; remove return, remove tab, reformat paragraph, curse, repeat...

Come 2019, and the Covid pandemic, I was determined to spend a few sunny days in the garden paying the manuscript the attention which it clearly deserved.

I eventually got around to it over the following Christmas.

I was amazed! Yes, I knew a little about the countries, and I'd certainly previous heard quite a lot of mutterings about the politics and problems, particularly as regards one specific country's natives (which shall here remain wholly un-named, but who clearly left such a negative impression that they coloured our many family geopolitical arguments and discussions over each Christmas, Easter, and just about every other meal-cum-debate thereafter). But what I'd never known, I realised, was the actual man described therein. Well, wrong: on one level, none of it surprised me at all. It was nothing but utterly typical of his upbringing, his background, his generation. But I'd definitely never heard of all these exploits. I wondered, in fact, if Mum ever had: though what went on between the main pages of those frequently-exchanged airmail letters I'll never know. I was only allowed to fill up, and in return read, the last, rear section, and never anything else.

Come the next Covid 'lockdown' - or was it the one after that? - I was seriously determined to get it edited it into a publishable form.

I finally got around to starting it in mid 2022. And then, I decided to move house, and other things got in the way and took up time, and... Sorry again, Dad.

But here it finally is. Quite a few bits - sometimes annoyingly-important bits - didn't fully survive the transition through the various word processing programmes. Several others were left either unclear, contradictory, or, in some aspect or another, just seemed downright wrong. I've tried to research, check, and clarify where I can. I've had to dodge around a few dates and facts which I don't think can be correct. I've added in a snippet or two - informed either by my own memories, from actually having been there with him, or from various other work-related documents and publications which he'd kept - and I've altered or deleted the odd thing (but not too many) which possibly might not go down at all well in our current times. The rest is much as he wrote it; and, as far as I've always trusted my father, true.

I hope that it meets with his approval.